Sail Sma

Sail Smart

MARK CHISNELL

Registered Office
John Wiley & Sons Ltd, The Atrium, Southern Gate, Chichester, West Sussex, PO19 8SQ, United Kingdom

For details of our global editorial offices, for customer services and for information about how to apply for permission to reuse the copyright material in this book please see our website at www.wiley.com.

Library of Congress Cataloging-in-Publication Data:

Chisnell, Mark
Sail Smart | Mark Chisnell
p. cm.
ISBN 978-1-119-94237-5 (pbk)
1. Sailing. I. Title
GV811.C515 2012
797.124—dc23
2012037467

A catalogue record for this book is available from the British Library.

ISBN 9781119942375 (pbk); ISBN 9781118337356 (ebk);
ISBN 9781118337349 (ebk); ISBN 9781118337332 (ebk)

Cover images: Front cover courtesy of B&G

Set in 11/13pt Gill Sans Std by MPS Limited, Chennai, India.
Printed by Printer Trento, Italy

Contents

Introduction

A change in the Racing Rules of Sailing in the mid 1980s led to performance-related electronics being allowed on sailing boats for racing. Since then, advances in modern technology and a different type of racing have all helped to alter the navigator's job to the point where it is almost a misnomer. Not totally, because, if only for safety's sake, there will always be a need for the traditional skills of chart work and position fixing. But if we assume that the equipment remains working then 'navigation' in a world where your position can be shown to within five metres on a colour chart screen sets a low bar. This is particularly true of the racing that is currently most popular – inshore Olympic style courses where water depth is no more relevant than to a dinghy sailor, and your position is only important in the context of the tactical situation. Today, the problem is how best to process the myriad of other data that comes out of the instrument system, and understand the story that it is telling us.

This is the problem to which this book is devoted. There are many excellent works on the more traditional aspects of navigation, and more are published all the time. So we shall leave to these others the intricacies of charts, tidal diamonds and dead reckoning plots and concentrate on what will most profit the modern race boat navigator – how to get the right information out of the electronic equipment to help the boat win races. In a sense, we will define a new role for the 'navigator'; a role that will enable him to contribute to success or failure on the race course in a lot more ways than was possible twenty years ago. Jobs like data collection and performance analysis were scarcely possible thirty years ago, but now make the navigator's role both more interesting and valuable.

We will be concentrating on the equipment normally used in the mainstream of yacht racing; position fixing and instrument systems. Whilst radars, autopilots and weather satellite systems all have parts to play in specialised races, they are of little direct consequence to those of us who do not wish, or have no opportunity, to race round the world or across oceans. Of these two the instrument system provides the greatest challenge. Everyone on-board a racing yacht will use the instrument system at some time or another. Tacticians for wind information, helmsman and trimmers for boat speed, mastmen for time to the next sail change. But someone must be responsible for this information, its accuracy, collection, assimilation, comparison to what has gone before and projection into what lies ahead. This is the job description of the navigator, and the contents of this book.

Position Fixing Systems

SOME THOUGHTS

Accurately fixing your position in relation to both the nearest land and the race course was, for a long time, the essence of being a good navigator. Knowing where you were as you hurtled towards a lee-shore finish line in thirty-five knots – with the 1.5 oz up, and a night whose blackness is only relieved by the luminescence of the white horses – was no joke before the advent of electronic position fixing systems. In the days of the Global Positioning System (GPS) the problem is almost, but not quite, trivial. The skills of good position fixing now mainly consist of finding the power switch.

Once you've got the GPS installed and powered up, as a navigator, you then want latitude and longitude, and some idea of its accuracy combined with a relatively easy means of converting this into something more useful; such as cross track error or speed over the ground. The advice in this book is, of necessity, general; the detail of what you need to know can only come from the manual. Whatever they say about it being the last resort, you should read it. But what we are really interested in here is how to use this information to help you win races, and this is the topic of the final section of the chapter.

THE GLOBAL POSITIONING SYSTEM

The Global Positioning System, or GPS as it is more commonly known, has become the system of choice for all sailors, replacing the likes of Decca, Loran and Omega. It provides continuous, global, all-weather navigation in three dimensions with high accuracy and great simplicity for the end user. Its accuracy is such that it can be used for much more than just position. Starting techniques and two-boat tuning have changed because distances – to the line and between boats – can be measured more accurately than they can be assessed by eye.

So how does it work? A receiver aboard the yacht uses timed radio signals to calculate distance from several transmitting beacons. The position of the beacons is known, and the consequent lines of position are used to triangulate the yacht's position. The radio beacons are aboard (see Diagram 1.1) a constellation of 24 satellites, which orbit the

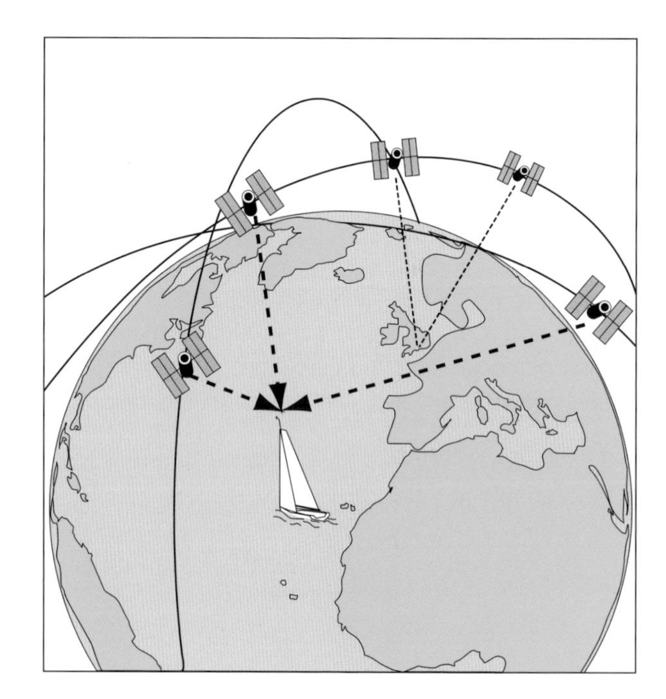

DIAGRAM 1.1 A ground control station monitors the orbit of the satellites and re-transmits corrections. Timing and orbit data is transmitted from the satellites to the yacht's receiver. Information from three satellites allows the yacht to calculate her position in two dimensions.

earth at a height of 10,900 miles. A minimum of four should be visible at any one time, which allows us to calculate position in three dimensions. Trigonometry tells us that each measurement places us on a sphere, and three of them intersecting will place us at two possible points. One of these solutions will be ridiculous, nowhere near the earth for instance, so it can be dismissed. The other point will be our position in three dimensions, latitude, longitude and altitude. To see why we need the fourth satellite we must look at how the distance measurements are made.

The distance measurement uses a concept called pseudo-random codes. A pseudo-random code is a succession of noughts and ones, apparently at random, but which actually repeats itself over a period of time. The satellite transmits these pseudo-random code messages, and the receiver listens and compares it to its own internally generated version of the code. All the satellites work on the same two frequencies and they are identified by having their own codes. The underlying assumption is that the code was issued from both sources simultaneously and so by comparing the codes, the receiver can calculate the time that the signal took to reach it.

The critical aspect of all this is the timing, due to the speed light travels the tiniest timing error will lead to quite big distance errors. So when we say the codes are issued simultaneously from the satellite and the receiver, it must be simultaneous. This requires the most accurate possible clocks on the satellite and the receiver. Putting $100,000 atomic clocks on 24 satellites is one thing, but if every user set required one the potential market would shrink somewhat! The solution is to make an extra distance measurement. We know that the clock error will be consistent to all three measurements. If this is the case then if we introduce an extra measurement the lines of position will not meet at a point (Diagram 1.2). We tell our receiver that if this happens then it must assume it has a clock error and adjust all its measurements by the same amount until they do meet at a point.

In this way we can eliminate timing errors in the receiver. Timing errors in the atomic clocks aboard the satellites do exist, and may occasionally produce a small unknown error in our position even though they are checked and corrected by the ground control system. Of the other errors in the system the worst are probably those due to the ionosphere, which slows down the GPS radio waves by an unpredictable amount. We can take account of it using 'average' figures for the effect, but obviously this will never be absolutely right. However, it may not be such a bad solution in the marine case because we only require latitude and longitude, and ionospheric error mainly affects vertical position and time.

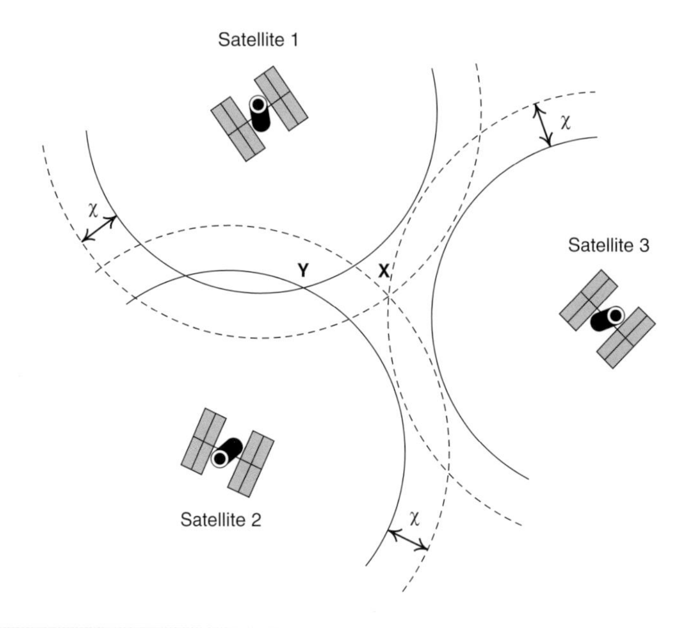

DIAGRAM 1.2 Clock Error. Given only satellites 1 and 2, the continuous lines which include the receiver's clock error meet at point Y. If we then add satellite 3 they no longer all meet. The receiver assumes that this is due to clock error and that the error (χ) is the same for all measurements. It adjusts them by the same amount until they meet at point X – the receiver's clock error corrected position.

The atmosphere does have some effect on GPS signals. Water vapour will slow and absorb them and unfortunately it is not an error that we can do much about. One that we can resolve is multi-path: the reflection of signals from other surfaces before they reach the receiver, which gives erroneous distance measurements. This can be almost eliminated by continuous tracking of the signal – but it is only something that the more expensive receivers will be likely to do.

When we look at uncertainties in the measurement we must include in our analysis Geometric Dilution of Precision, or GDOP. It effectively magnifies the other errors, depending on the angle between the satellites and the receiver; the wider the angle between the satellites that you are using, the bigger the GDOP and the bigger the overall error. The best receivers will look at all the satellites available and choose those that will reduce the GDOP as much as possible – or simply use all of them. GDOP is usually quoted as a single number; you might expect something between two at the best and ten as the outer limit of acceptability.

Initially, the highest quality GPS signal was reserved for use by the military whilst civilian GPS units could only receive a signal that was intentionally degraded, this was called Selective Availability. In 2000, Bill Clinton signed an executive order turning off Selective Availablity at midnight on 1 May 2000. This resulted in the improvement of the precision of civilian GPS units from 100 metres to 20 metres.

If you want better than 20 metres – and you will for start-line calculations or two-boat testing – then one way of getting it is Differential GPS. This corrects the GPS signal by having an additional ground station in a known position. It can then calculate the error in the GPS position at that point. Because the satellites are so high up this error will be the same in the area around it and the correction can be transmitted to other receivers and used to calculate their positions precisely – and precise means within two to three metres.

Another development in positioning technology that can provide greater accuracy than the standard GPS is the Wide Area Augmentation System (WAAS) in the USA, and the European Geostationary Navigation Overlay Service (EGNOS) in Europe. Both systems were developed to assist air navigation by improving accuracy, integrity and availability. The initial goal of WAAS was to enable aircraft to rely on positioning from GPS for all elements of flight, including precision approaches to landing.

Like Differential GPS, WAAS and EGNOS use a network of ground stations to measure the small variations in the GPS signals. The data from these ground stations is then collated at master stations and corrections are sent to geostationary WAAS or EGNOS satellites, where they can be beamed back down to WAAS/EGNOS enabled GPS antennas.

Typically WAAS/EGNOS enabled GPS antennas provide an accuracy of less than two metres, though the coverage is not global. It is this kind of precision that was costing America's Cup teams thousands of dollars at the start of the new millennium, but within a decade had become so affordable that it's standard in most entry level GPS receivers.

Equipping yourself with the best GPS possible will go a long way to help with any testing that you may embark on. In the 2000 America's Cup advances in Differential GPS allowed highly effective two-boat testing to take place, with two identical boats testing sails, spars and foils against each other. Instruments systems have long struggled to measure comparative performance to the kind of accuracy required to split two boats of race winning speed.

Two-boat testing in the America's Cup is no different in principle from dinghy testing – you sail the boats beside each other and see who is going faster. But this is not as simple as it sounds, if the wind heads the leeward boat will seem to get an advantage, if it lifts the windward boat will look good. Whoever is judging the test must not only assess the relative distance and angle of the boats, perhaps using a hand bearing compass or stadiometer, but they must also take account of any wind shifts. It takes practice, and even then requires a great deal of concentration not to make mistakes and come to false conclusions – conclusions that might send your whole design program up the wrong path.

It's different with a Differential or WAAS/EGNOS enabled GPS, a computer running tactical software and a couple of radios. Using this equipment it is possible for anyone to begin to make valid comparisons. It can be taken a step further by transmitting the wind data and the position of the boats via telemetry to a computer; and then it is possible to resolve the distance apart with respect to the wind. At the same time, differences in wind strength or direction between the two boats can be spotted. Given suitable hardware the computer can record the results of each test run for you.

Another consideration for any serious racing team is to get a GPS that runs at greater than the standard 1Hz update rate (one update per second) – a 5Hz update rate (5 times per second) would be a significant improvement. This becomes important in the starting sequence when a 5Hz unit is able to keep up with the rapid changes of course. When it's interfaced with tactical racing software it also helps to provide a much better picture of the start. A quick updating Differential or WAAS/EGNOS enabled GPS will call the position of the yacht relative to the start line more accurately than all but the very best bowman or woman!

USING THE GPS

The basic task of a GPS is to tell you where you are, traditionally done by displaying a latitude and longitude. This obviously has its uses, placing you on a chart or tidal atlas in relation to the shore, its effects and the tidal streams. But more often you need to know where you are in relation to the next mark. Most GPSs have ways of telling you this as well as other numbers, but exactly what data you need and how you use it depends on the type of leg you are sailing. What we will look at next is the information you can get from a GPS, and ways – and ways not – to use it.

COG and SOG; Course and Speed over the Ground

Those of us who race in tidal waters will be familiar with COG and SOG. If the water is stationary relative to the ground then your boat speed and heading tell you all you need to know about your motion relative to the mark. But if the water is moving, either through tide or current, then your course and speed across the ground will be the vector addition of your boat's motion through the water, and the motion of the water relative to the ground (Diagram 1.3). This is fine if you know the exact rate and direction of the current, but that's unlikely unless you have been able to sit by a fixed buoy and measure it, or have an extremely accurate tidal atlas.

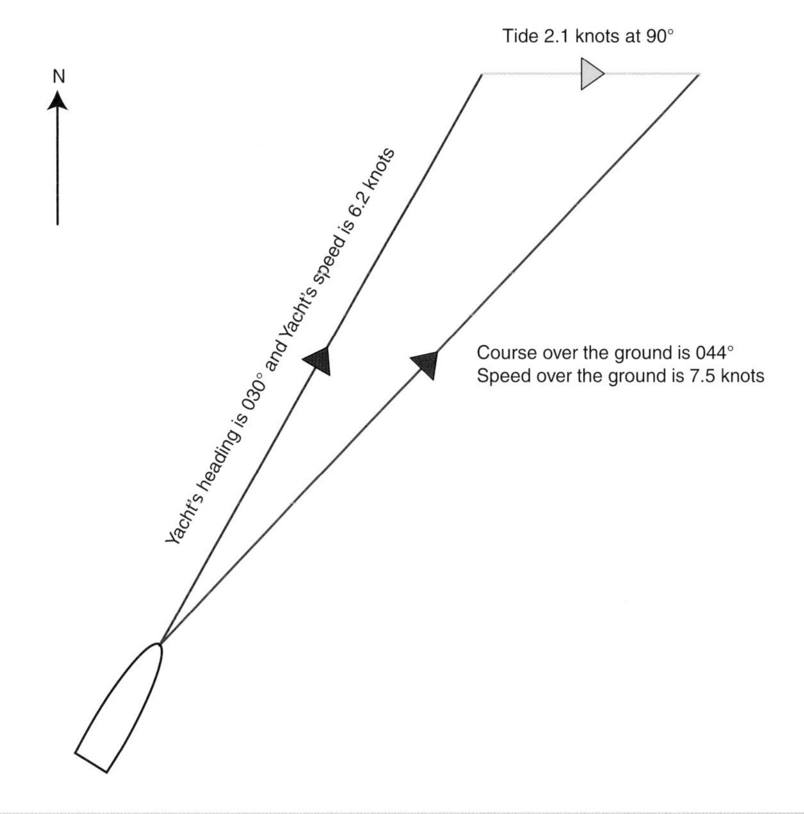

DIAGRAM 1.3 COG and SOG. The yacht's motion through the water is added to the water's motion across the ground to give the yacht's speed and course over the ground.

It would be far more useful if we could do the calculation the other way around. If we knew course and speed across the ground then we could subtract our boat speed and heading through the water, and the result would be the direction and speed of the tide/current. This is one use of the COG and SOG figures produced by the GPS, they allow you to calculate the tide/current you are sailing in and then, hopefully, turn it to your advantage. Some instrument systems, if interfaced to the GPS, will calculate tide/current for you. The accuracy with which this can be done depends on the quality of the GPS. But a decent unit will allow you to take advantage of very subtle differences – dodging out of a couple of tenths more Solent tide is quite possible these days.

Using COG and SOG to dodge unfavourable tides is straightforward, but tidal strategy gets more subtle. Take as an example a reaching leg that is sufficiently long for the tide to change significantly, in rate or direction, during the leg. A situation that is different from one where the leg is so short that the tide is more or less constant. In both cases the shortest course is the one which you can sail on a steady compass bearing. Now if the tide is constant the COG will also be a steady bearing, but if the leg is long enough for the tide to change, the COG will vary whilst your steering bearing will be the same. Trying to keep the COG the same on a reaching leg with a changing tide is wrong. In a constant tide, we are trying to achieve a heading that will compensate for the effect of the tide and allow us to sail on both a steady steering bearing and a steady COG. The COG should be the direct line across the ground between the marks. The steering bearing can be calculated if you have an estimate of the tide (Diagram 1.4). But in these days of on-deck navigation, or if you do not know the exact strength and rate of the tide, then the COG can be used as a 'cheat'. You just sail the boat at whatever steering bearing gives you the necessary COG to the mark.

But this technique is disastrous if you try to employ it on a leg where the tide is changing. No sooner have you settled on a steering bearing that gives you the right COG than the tide changes and you find the COG changing with it. So you correct the steering bearing again to get the COG right and off you go. But as the tide changes the COG will alter once more, and so you will have to change your steering bearing again. This cycle continues with you steering dog-legs across the ocean – not fast at all.

What you need to calculate is the total amount of tide for the time you are on the leg. Then you can set a course that will compensate for the net effect of the tide over the whole leg. Your calculation will start with an estimate of your speed and the distance to the mark to work out how long you will be on the leg. Then, using tidal data you can work out how far you will be pushed off the rhumb line by the tide during each of those hours. Adding them all together you will be left with a net amount of tide pushing you

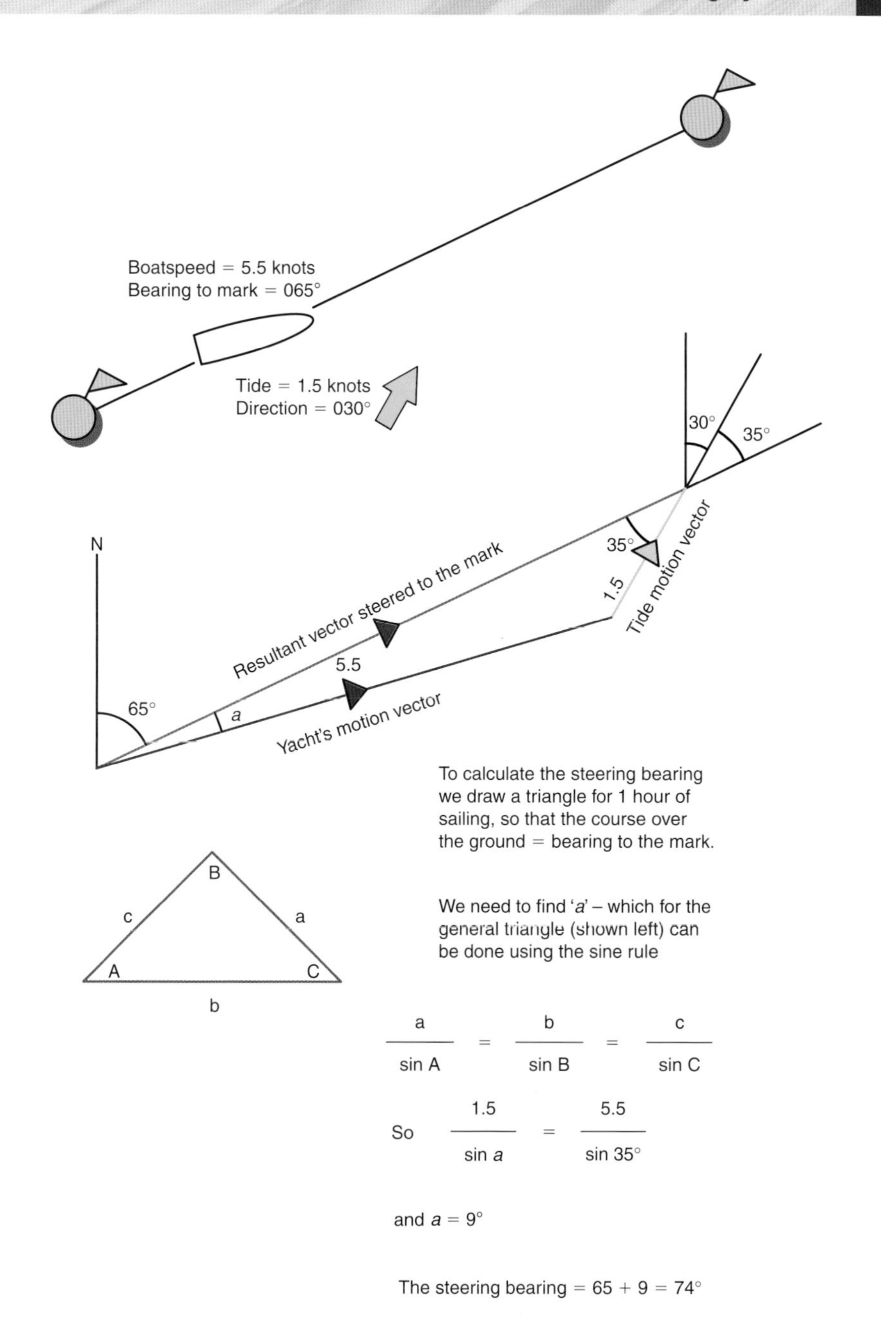

$$\frac{a}{\sin A} = \frac{b}{\sin B} = \frac{c}{\sin C}$$

$$\text{So} \quad \frac{1.5}{\sin a} = \frac{5.5}{\sin 35°}$$

and $a = 9°$

The steering bearing = 65 + 9 = 74°

DIAGRAM 1.4 Calculation of a steering bearing for a course affected by constant tide.

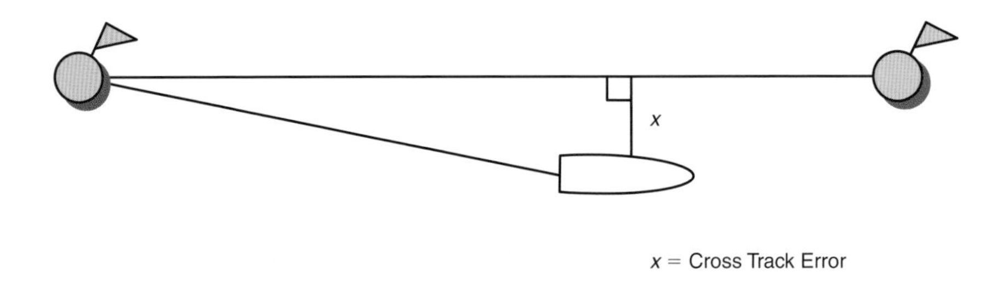

DIAGRAM 1.5 Cross Track Error is the perpendicular distance from the straight line to the mark.

one way or the other – you then calculate a tidal offset to account for just this much tide in the same way as you would for a constant tide. You should be able to sail on the resultant bearing for the whole leg and end up on the mark. Of course, it doesn't usually work like that, if the wind varies in strength or direction the time you spend on the leg will change, and the tidal estimates are rarely that accurate. So the calculation is one that you must continually repeat as you sail down the leg.

Our main interest here is how the instruments can help us to sail a leg like this, and this is a good time to introduce Cross Track Error, a GPS function that can be particularly useful on this type of leg. Cross Track Error is your perpendicular distance from the straight line course between the last waypoint and the one you are sailing to (Diagram 1.5). For the GPS to be able to calculate this you will need to have programmed in the relevant waypoints. Its use in this instance is that it enables you to keep a track of exactly how much you are getting swept off the rhumb line as you sail down the leg. If after two hours you reckoned you would be half a mile south of the rhumb line, and the Cross Track Error only puts you a quarter mile south, then it gives you plenty of time to try and work out what is happening and how to correct it – before you end up a quarter mile south of the buoy and beating back up to it in a foul tide. A move that is unlikely to endear you to any but the most relaxed of skippers.

So much for reaching, the next two GPS functions we will look at are often of more use on a beat or run. On these legs the navigator abdicates much of the responsibility, particularly on short courses, to the tactician. The COG, SOG and Cross Track Error are not directly relevant to the boats heading, though they still need to be considered. Positional information that the tactician needs to know is the range and bearing of the next mark, and hence your proximity to the two laylines (Diagram 1.6). Whatever

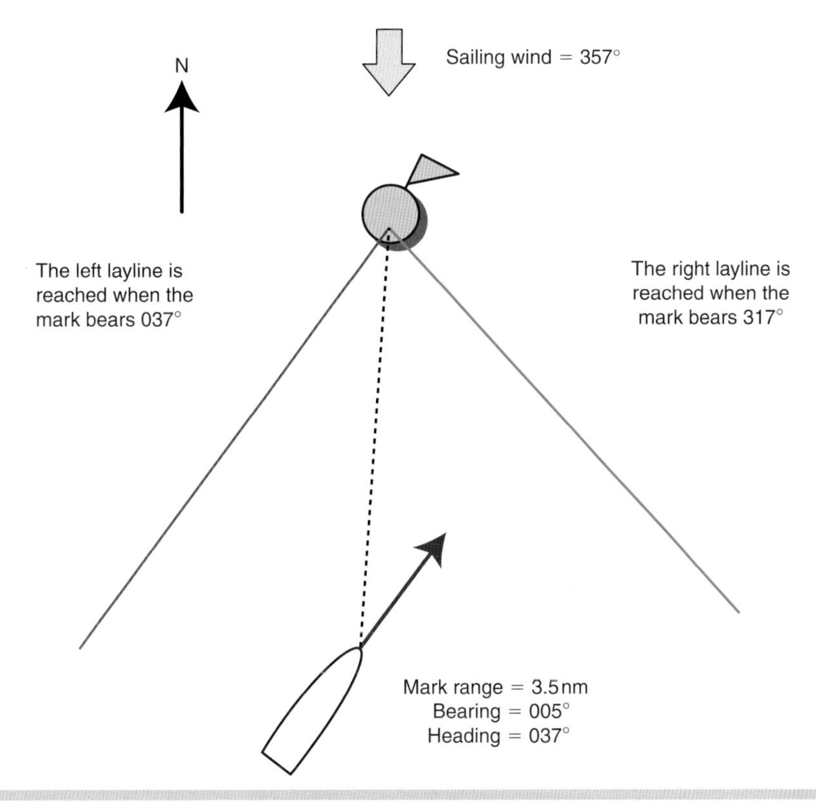

DIAGRAM 1.6 If you check your heading on each tack you can work out your tacking angle – in this case 80°. Then even if the wind shifts you will know what the mark must bear in order to lay it. If the mark is in your GPS, checking its bearing will tell you when to tack. When there is tide you can either work out how much to compensate in the normal way, or watch your COG to see how it is affecting you. Your COG on each tack will be the layline bearing – but check the wind, if it alters so will your COG.

tactical considerations he may have up the beat; wind, tide, shore effects or other boats, he will want to place the boat carefully on the course. Not getting too much to either side too early, and certainly not overstanding. You can use the range and bearing to the mark, together with a knowledge of your COG and SOG to give him this information.

There is one important use of COG or Cross Track Error on a windward leg. This example is when you are beating for several hours towards a mark whose bearing lies approximately perpendicular to the tidal stream. In that time you may have several hours of tide running left to right, then several hours of tide running right to left. It is a basic strategy of this type of leg that you set off on the tack which puts the tide under your leebow. I do not want to get into the leebow effect here, mainly because it would pre-empt what I want to say in the next section, but there is no question that if you have a tide running across a beat, the tack that puts the tide under your leebow will take you a lot closer to the mark than the one which puts it on the weather bow (Diagram 1.7). The

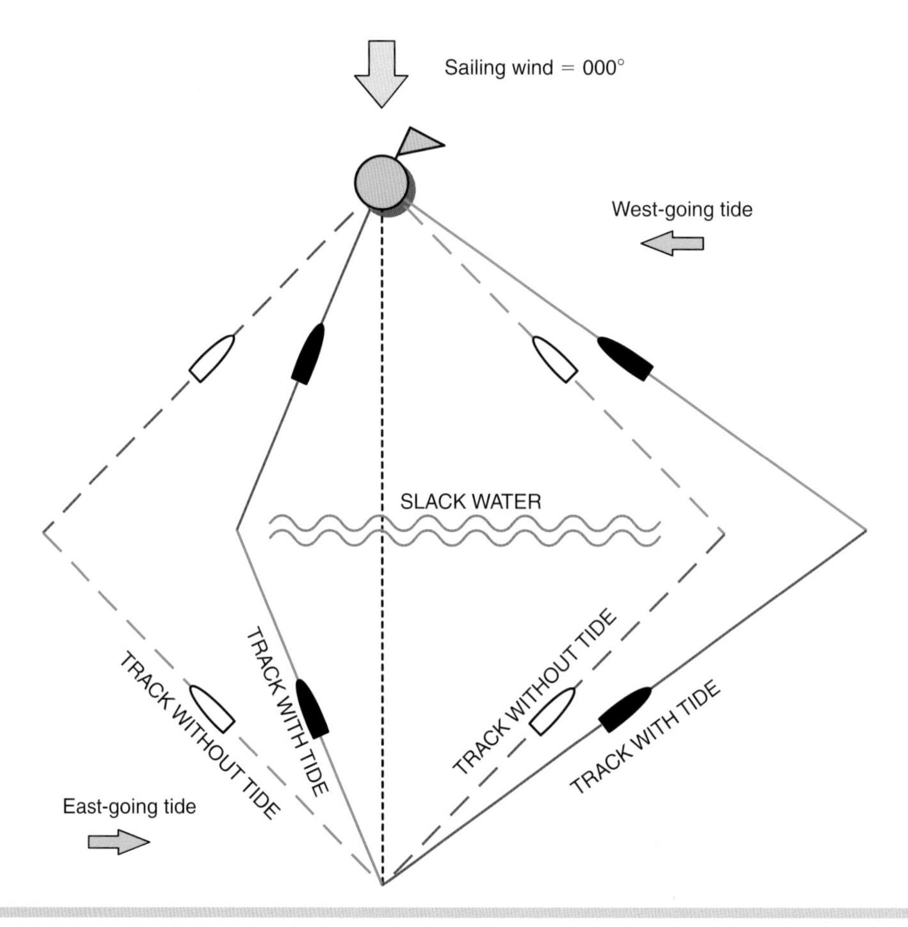

DIAGRAM 1.7 Sailing with the tide under the lee bow keeps the boat closer to the rhumb line (and optimises VMC). COG and Cross Track Error can be used to work out when the tide is changing, which is also the time to tack. Other advantages of this technique follow in the next section.

beauty of a long beat where the tide changes half way up is that you can then flop over onto the other tack and so spend the entire time with the tide under your leebow. The advantage of doing this is so great that you would have to have some extraordinarily good reason for not doing it – such as a big, and guaranteed, wind shift!

So where does COG and Cross Track Error come into this? Well, the key to it is spotting when the tide has changed and the other tack now has the leebow. Cross Track error will tell you this quickly. When the tide is on the leebow it will be pushing you up onto the rhumb line, and so the Cross Track Error is a lot smaller than it would be without the tide. As soon as the tide changes the Cross Track Error starts to shoot up as the tide is then taking you away from the rhumb line. The speed with which it increases is your indicator on when to tack. Similarly, with COG it will be steady until the tide goes round, when you see it start to alter then it is time to go.

Unfortunately we rarely get a windward leg with steady wind and a tidal change exactly half way along it, and even if we did there may well be some other complicating factors at the next mark, such as a shoreline effect. But what this technique of leebowing does do is maximise your speed towards the mark – in the jargon: it optimises your VMC. Whether or not this is a good general strategy is something we are going to discuss in the final chapter on instrument techniques using the polar table.

Instrument Systems

STAND-ALONE EQUIPMENT VERSUS INTEGRATED SYSTEMS

Yacht instruments can be divided into two types; the first type are stand-alone units that provide one piece of data each; the second type is more fittingly called a system – the separate devices being integrated into a whole that is greater than the sum of the parts. Much of what we say here will be about integrated instrument systems, the reason being that the results obtained from a stand-alone system are quite limited.

To understand why this is let us split up an instrument system into its constituent parts of which there are three:

- The sensors that actually measure the physical effects, be it boat speed, wind angle or compass heading;
- The software that translates the raw sensor data into a number we can understand;
- The display unit that communicates this number to the world.

A stand-alone device, however, consists of only the following:

- A single sensor, some software and a display.

It is not connected in any way to any of the other devices on the boat and it is solely dedicated to the task of telling you one of a limited number of things; the boat speed, heading, depth, wind speed and angle across the deck; or perhaps the load on various parts of the rig like the headstay. All of this information is useful, but not that useful. Instead of just knowing the speed of the wind across the deck would it not be better if we knew the actual speed of the wind across the water? We cannot measure this, but we can calculate it using the boat speed, and the angle and speed of the wind across the deck.

This however is what an integrated instrument system can do; it takes the raw data from all the different sensors and then uses a vector diagram called the wind triangle (see next section) to calculate the numbers that will help us race the boat. In addition, because all the numbers are calculated in the same box they can be sent out to any display connected to it. This is a lot more powerful. Not only do you have more information, but you can display it wherever you want rather than being locked into just one display for one piece of data as you would be with a stand-alone system. For effective race boat instruments an integrated system really is the only choice.

THE WIND TRIANGLE AND SOME NOMENCLATURE

As we discussed in the section above, an integrated instrument system uses data from several different sensors then applies some maths to the numbers in order to calculate additional information. The maths needed for this is known as the Wind Triangle, a vector calculation which we are going to look at here.

The four measurements that are made are:

- The boat speed, which comes from some form of impeller, paddle wheel or solid state device and is exactly what it says it is – the speed of the boat moving through the water;
- Next is the compass heading, which is straight-forward – the magnetic heading of the boat;
- The final two are the apparent wind speed and the apparent wind angle, and here we need a definition. The 'apparent wind' is the breeze that you feel blowing across

the boat, i.e. the one that you can directly measure on-board. It is the product of three components:

- The wind blowing across the land – which we call the ground wind – is the wind that you will see on the weather maps.
- The wind produced by the motion of the water relative to the land – which we call the tide wind – is equal in strength and opposite in direction to the water flow.
- The wind produced by the motion of the boat relative to the water – which we call the motion wind – is equal in velocity to the boat speed and blows directly onto the bow.

- The vector product of all three is the apparent wind, which is the only one of these that we can measure directly. The wind triangle allows us to calculate the rest.

Let us assume for the moment that we have no GPS connected into the instruments and that we are out of sight of the shore; we have no idea whether the water is moving relative to the land or not. The wind triangle uses the boat speed, apparent wind speed and the apparent wind angle to calculate what we are going to call the true wind angle and speed – the third side of the triangle (Diagram 2.1). These two

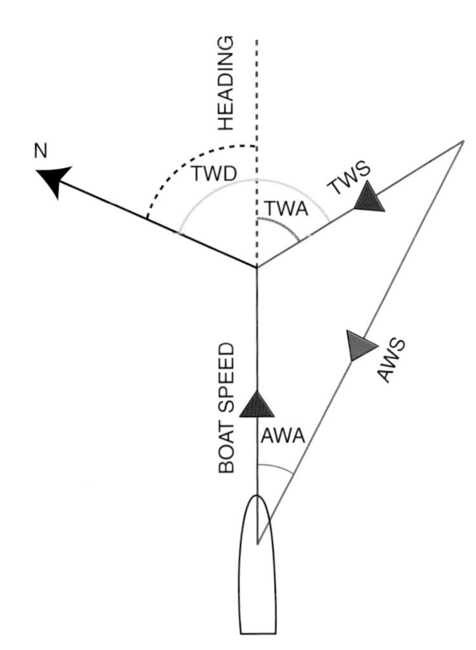

DIAGRAM 2.1 The wind triangle connects Heading Boat Speed, Apparent Wind Angle (AWA), Apparent Wind Speed (AWS), True Wind Angle (TWA), True Wind Speed (TWS), and True Wind Direction (TWD).

numbers are a great deal more useful than the apparent wind speed and angle, for lots of reasons that we will see as we continue with the book.

The point I want to stress now is what happens when we relate this true wind angle we have calculated to the fourth measurement that we have made, which is the compass heading. The result that we get is what the instruments usually call the true or magnetic wind direction. This is probably the most useful tactical tool that you have on the boat and because it is calculated independently of the boat's direction it is the most precise measurement you have of the wind shifts that you should be using to sail quickly round the course.

The term I prefer for this is the sailing wind, because it is the vector sum of the ground wind and the tide wind and it is the wind that we actually sail in (Diagram 2.2). The most important point to grasp is that the changes in wind speed and direction that you see on the sailing (or true/magnetic) wind on the instruments are not just caused by the ground wind altering; they are also affected by the tide changing, because of the tide wind component. The consequences of this fact cannot be underestimated. We will start with an example we used earlier in the book, where we were talking about a cross tide beat which is where the tide changes. The reason for always being on the tack where the tide is under your leebow is because this tack is lifted due to the tide

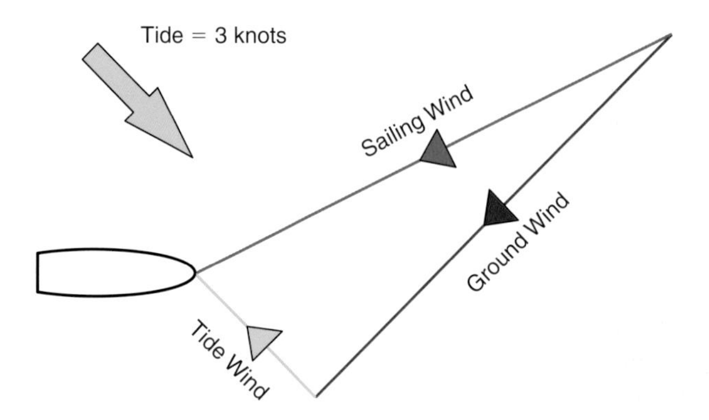

DIAGRAM 2.2 The ground wind and the tide wind combine to form the true or sailing wind, whose direction and speed can be calculated from the boat speed, compass, apparent wind speed and angle, as above.

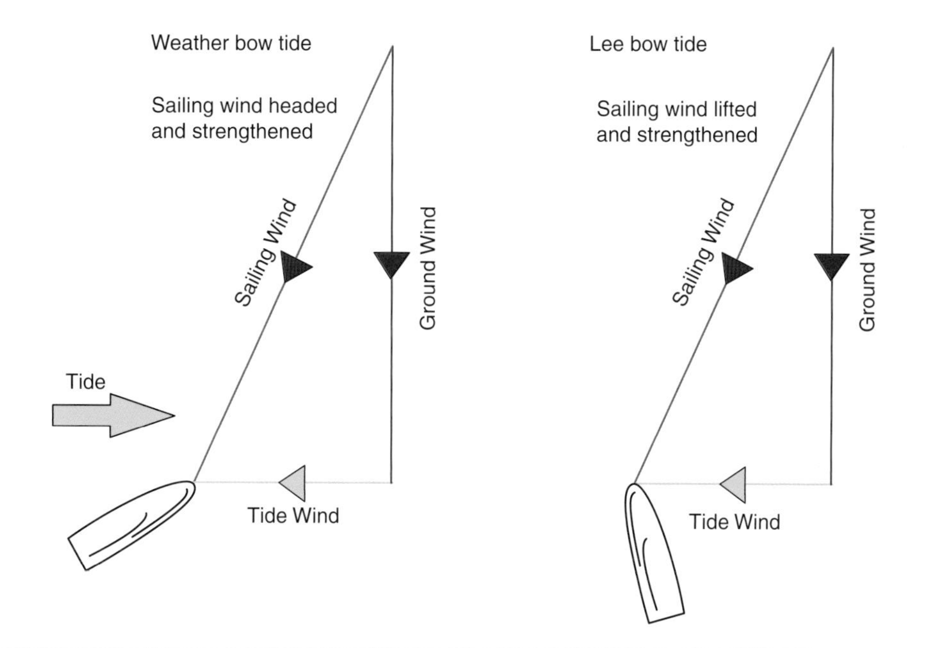

DIAGRAM 2.3 The real lee bow effect.

wind component. If you sail on the other tack, then you are sailing on a header that will disappear when the tide changes (Diagram 2.3). This is why you tack when the tide goes round – so you are always sailing on the lift in the sailing wind created by the tide.

One place that you see a lot of tidal dependence of the sailing wind is the Solent (Diagram 2.4). A typical situation would be sailing upwind beside the shore whilst dodging an adverse tide; as you sail out of the shallow water and into the stronger tide on port tack, the introduction of the tide wind component will cause the sailing wind on the instruments to head by ten degrees. Watch out for the jib lifting and the helm coming down as the helmsman accounts for the header. Then when you sail back into the shore on starboard and out of the adverse tide you will be lifted again. This is why it is so important to distinguish between ground and sailing winds – if the wind effect you saw on the last lap was caused by the tide wind, and the tide has now changed, then it is no good looking for the same effect again. Equally, if you know the

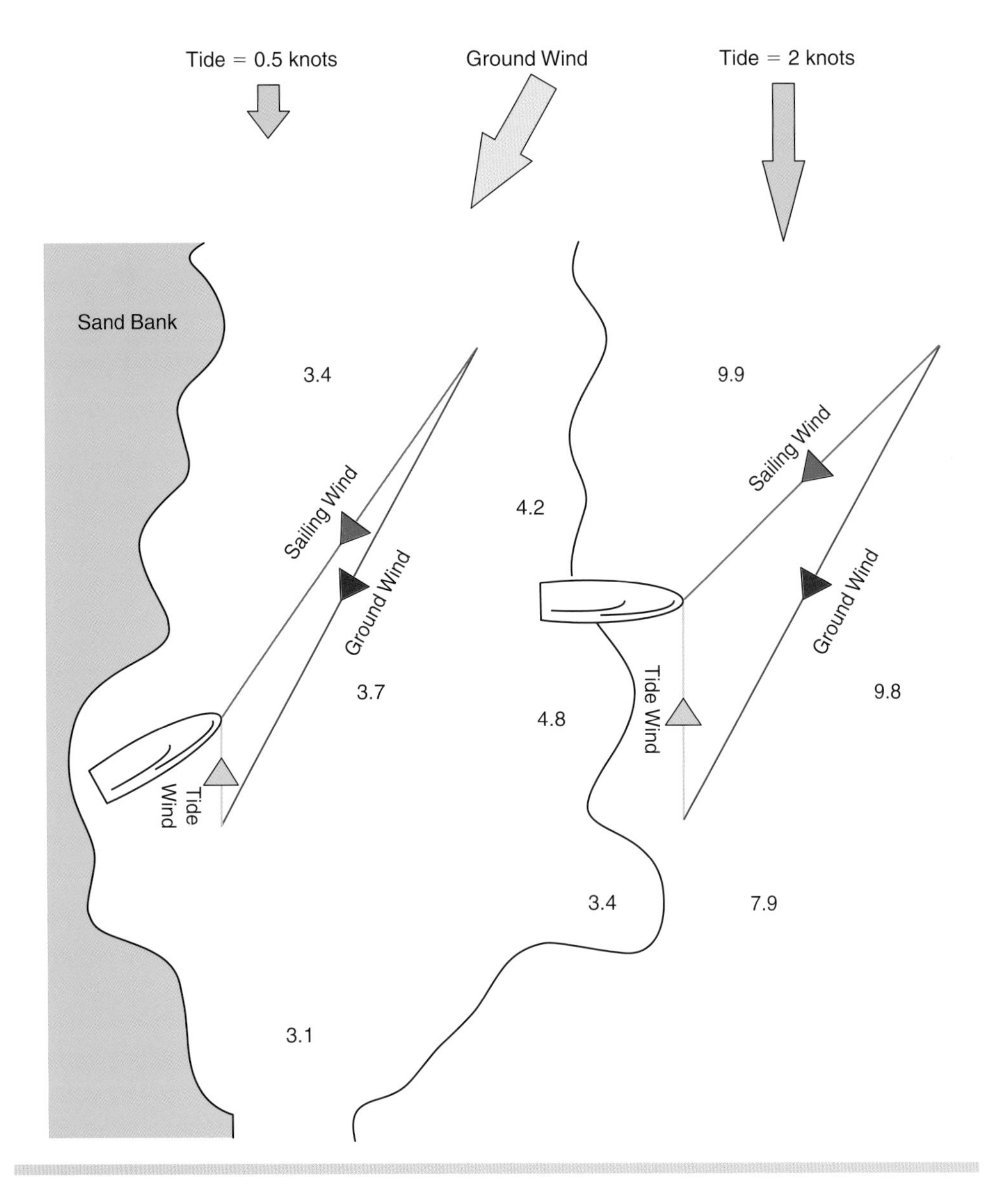

DIAGRAM 2.4 The sailing wind is both heading in direction and decreased in speed as you sail into the stronger weather bow tide.

tide will change while you are on an upwind or downwind leg you need to get on the lifted tack straightaway, before the tide changes and the lift disappears.

Much the same can be said of what is called the true wind speed (true wind speed is the more common name, although it would probably be more accurate to call it sailing wind speed). The true wind speed will go up and down with the tide. If the tide wind is increasing your true (sailing) wind speed then use it tactically before the tide changes, to cross a patch of rough water perhaps. To achieve this sort of tactical thinking you will need to know what the tide is doing so that you can work out the tidal component of the sailing wind you are seeing on your instrument dial. As well as doing the necessary preparation with a chart or tidal atlas you also need to watch the tide on buoys, and of course the values of COG and SOG as compared to your boat speed and heading, which will also tell you how much of the wind is tidal.

Setting Up an Instrument System

SOME COMMENTS ON THE CALIBRATION PROCESS

Three things should be said about the calibration of yacht instruments. Firstly, the importance of doing it at all. Any instrument system is only as good as its calibration. You can spend a fortune on the best equipment, but if you do not get out there and set it up properly you might as well not bother. In fact you are probably worse off than you would be with just a compass on board.

My first experience with instruments was aboard 12m's in the 1987 America's Cup in Fremantle. After a startling promotion from compound sweeper to navigator, I stepped aboard for the first day's sailing. Scarcely had I switched the instruments on than we were in the middle of a major down-speed tacking duel, by which I mean that the boat is spun back into the next tack before it has picked up the lost speed from the last one. There seemed to be a lot to do: winding runners, watching layline proximity and checking the shifts. I decided to concentrate on the wind so that I could at least quickly answer the most obvious question 'are we up or down?'. After five or six tacks, I had come to the conclusion that the wind direction I was watching on the dial bore about as much relevance to the wind blowing over the water, as it did to the odds of us winning the America's Cup. The result was that with several tens of thousands of pounds worth of equipment at my disposal I was completely unable to work out whether we were headed or lifted – not an impressive start.

What slowly became clear was that the wind direction was different on each tack, not only that but the size of the error varied with the wind speed, and finally that the system was set-up with such a value of damping that in a down speed tack the instruments were not settling to the correct value before we tacked back anyway. All these are things that can be corrected by good preparation – so don't skip on it.

Secondly, you must approach the calibration in a very systematic order. Everything in the system is interdependent and if you start calibrating in the wrong order you are wasting your time. To see why this is so, we should look at the wind triangle we discussed in the last section. We can see that much of the key information provided by the instruments, such as the sailing wind data, is derived from the measurement of only four values: the boat speed, the compass heading and the apparent wind speed and angle. In fact there is a fifth, the heel angle, but how this is involved we will leave till later. If you start to calibrate one of the functions that is calculated from these numbers, such as the true wind angle, before you have calibrated all of its constituent measured values, like the boat speed; then subsequently calibrating the boat speed will upset all the work you have done on the true wind angle. So the rule is: calibrate every sensor that measures something directly on the boat first, and only when you are completely happy do you move onto whatever calibrations are provided for the other functions.

The phrase 'whatever calibrations are provided for the other functions' brings up an interesting limitation of this book. That the book must talk in general terms about something that is specific to the individual reader, i.e. the instrument system itself. It is a limitation that is going to be most apparent in this section on calibration. Although instrument systems provide the same data, advice on how you use it can be general, but the calibration arrangements are specific. For the values measured by the sensors this is not too much of a problem, the manual will tell you which buttons to press and there are plenty of general points to be made about what you are trying to achieve and the pitfalls to avoid. The problem arises most acutely with respect to the functions calculated by the wind triangle. Some of you will have systems that provide no calibration facilities for these numbers at all, others will have functions such as mast twist and upwash, or perhaps direct control over the true wind data itself. Ultimately though, whatever the system, the problem of calibrating it is the same. My hope is that explaining the problem will help you approach 'whatever calibrations are provided for the other functions' with a clearer view of what you are trying to achieve with them, and how they are supposed to help.

The final point is that calibration is not a one-time task, you never finish the job. The main reason for this is the phenomena of wind sheer and gradient, which we will discuss in the section on apparent wind speed and angle. Don't forget that all the sensors on the boat are mechanical devices that can be moved, bashed, jarred or kicked, from one week to the next. The instruments are never completely above suspicion, but equally when you are confident they are right you can learn from even the strangest readings – as we will see.

CALIBRATION OF THE COMPASS

It is always dangerous to generalise, but it's probably safe to say that most instrument systems 'generally' use fluxgate compasses. There are some great new compass technologies available, such as MEMS and Fibre Optic Gyros, but pricing keeps these sensors beyond the reach of all but the best funded America's Cup teams and Volvo Ocean Race teams. The fluxgate compass measures the field strength around it and so detects the direction of north. This signal is much easier to convert and input into a digital instrument system than the direction of the swinging needle of a conventional compass: that's the reason it has such widespread use in instrument systems.

Until the late eighties compass calibration was done by a specialist. Someone would come and tell you, at least on most stripped out racing boats, that you were a degree or two out here and there, and then usually leave it at that. Technology has intervened and provided much greater accuracy. The innovation I have in mind is the auto-swing facility, which allows the compass, if certain conditions are met, to calculate its own deviation card and then correct the errors out.

The way these compasses work is to measure the total field strength surrounding them as they turn through a circle. The total field strength measured has two components, one is due to the earth, and the other is from the magnetic fields within the boat – all the magnetic sources that create deviation. These two fields are fixed relative to different points of reference. The earth's field is fixed relative to the earth and so the yacht rotates within it. Whereas the deviating fields are fixed to the yacht and therefore rotate with the compass as the yacht turns. Because of this difference the compass is able to separate them out and so calculate the deviating field. Once it has done this, it applies the necessary corrections at all points of the compass.

To make all this happen you usually have to turn the boat in a steady circle, so that the compass can make its measurements, calculations and corrections. This is perhaps a great deal simpler than it sounds. There are many stories of people struggling to swing compasses, trying on calm days with no sea state and still failing. In one particular instance we managed to swing the compass under engine without any issues, but as soon as we went sailing the compass heading would lift by 10 degrees. We looked at where the compass was installed; low down, on the centre line in the owner's cabin. There were no high voltage cables, tool bags or radios around it, and yet we still had this problem. The heading when motoring was perfect, but under sail it was out. After much head scratching we concluded the only difference was that the propshaft was turning. When we looked on the yacht's drawings we found that the compass was mounted only 50 cm from the turning propshaft. We reinstalled the compass further away from the propshaft, swung it again, and lo and behold everything was fine.

The problem had been caused by the magnetic field of the spinning propeller shaft creating a constant disturbance in the compass during calibration. As soon as the engine was out of gear and the propeller stopped turning the magnetic disturbance disappeared. The lesson was that auto-swing compasses will auto-swing, but only up to a point – the rules about large magnetic fields being placed near them still apply.

CALIBRATION OF THE BOAT SPEED

Today, most boats are equipped with paddlewheel sensors. These are mechanical devices that rotate at a frequency proportional to the speed of the water-flow past them. If they are properly maintained they are extremely reliable; however, they are susceptible to fouling and should be cleaned regularly, then carefully replaced to ensure the sensor lines up with the centre line.

Boat speed calibration should be one of the first jobs for any new boat owner, or at the beginning of the season. If you allow the helmsman and trimmers to get used to one boat speed setting and then decide to re-calibrate they will find it difficult to adjust to the new readings. So it is important that you calibrate as early, and as accurately, as possible.

Every instrument system has its own methods for actually doing the calibration, be it just pressing buttons at the beginning and end of each run or using the trusty calculator.

Whatever the specific technique required for the instruments (which you will find in the manual – yes, read it!) there are some general rules that can be followed to get a more accurate result. Always steer as straight a line as possible between your chosen distance marks. If it is a proper measured mile then the chart will provide the bearing to steer between the transits. If you have worked out the run yourself then decide on what you are going to steer to and stick to it. If you waver from the straight line then the log will measure extra distance that will not be accounted for in the calibration calculation and give you an error.

You should choose a time when the water is flat. The log measures the water-flow past it and is not too choosy whether the flow is created by the boat moving forward, or up and down. If the boat is pitching it will record more distance than you have actually travelled. Pick a calm day, and, just as importantly, a time when there is not too much traffic about. Not only does the pitching from wash affect the calibration, but there is nothing more infuriating than having to alter course from your fixed bearing to dodge another boat half way through your final run, so ruining the entire calibration. Do the runs at fixed engine revs to keep the speed consistent. When you turn round at the end of the run the boat will slow down because of the braking effect of the turn. It is important that the turn is sufficiently wide so that the boat has speeded up again by the time you start the next run – otherwise the acceleration will affect the results.

CALIBRATION OF THE APPARENT WIND SPEED AND ANGLE

Apparent wind speed and angle are measured using a masthead unit that combines anemometer with wind vane. Their calibration problems are connected which is why we are dealing with them together. Development on the masthead unit has concentrated on reducing the weight and windage – critical at the top of the mast – and finding the right position in the air flowing past the boat. It was way back at the 1987 Fremantle America's Cup that masthead units started to appear in different places; going upwards, outwards, fore and aft from the masthead, and even being moved down to the hounds by the Italia syndicate. By the 2010 America's Cup, when we saw two multi-hulls racing each other, the wind vanes were located at the top of the mast and on poles at the back of either hull.

The idea behind all this is to move the measuring units to a position where they are least affected by aerodynamic errors, and can give the user the most relevant wind information. These errors are created by the deflection of the airflow along the sails (Diagram 3.1). We can see that the apparent wind is altered from that which is created by the pure motion of the boat in the sailing wind. The easiest way to understand this is to imagine a motor boat travelling beside a sailing boat at exactly the same speed and angle in a completely uniform wind. If you measured the apparent wind on the motor boat and on the sailing boat, using the instruments on a pole 10 ft off the deck, the results would be different. The difference, both in angle and in speed, is due to the deflection of the wind by the sails. Matters can be improved by putting the instruments on top of the mast and raising them as high as vertically possible to get them into clean airflow.

This has the downside that a physical error will be introduced to the angle measurement because the masthead unit twists when the mast tip twists. There is another physical error introduced by the heel of the boat, the angle and speed sensors are measuring the airflow whilst tilted at an angle to it. Although this can be corrected for with a relatively simple calculation, it is probably easier and no less accurate to lump it in with all the other errors involved and deal with them all together. Having done that, how do we go about correcting them?

There are two ways of approaching the problem for the apparent wind speed and angle. The first, which I favour, is to say that we will ignore these effects and accept the errors as part of the measurement. So we define the apparent wind angle and speed that we measure and use on a sailboat as including the deflection of that wind by the sail plan. Obviously this makes calibration of these numbers straightforward, but only because it

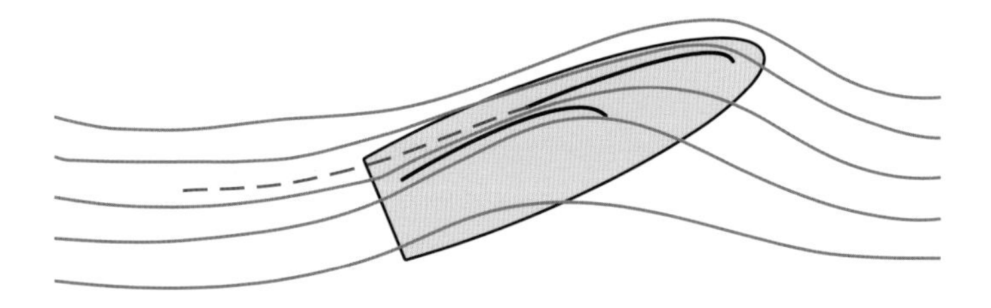

DIAGRAM 3.1 The sail plans effect on the apparent wind speed and angle can be seen here.

moves the problem down the line to the calibration of the true wind angle and the sailing wind direction. The advantage I believe this approach has is that the problem is more easily and simply dealt with when you are calibrating the sailing wind. The reason is that the errors caused by deflection of the wind by the sail plan show up much more readily in the sailing wind than they do in the apparent wind.

If you had a boat with just apparent wind speed and angle sensors you would never know that the apparent wind was anything other than correct. Your only real comparison point for the apparent wind angle is whether or not it is the same on both tacks, and once you have set this up there is little more that you can do. With the apparent wind speed there is even less, you have nothing to compare your measurement to – so you simply trust to the manufacturers and leave it. But as soon as you put on board the means of calculating the sailing wind then the defects become obvious. If the apparent wind is too narrow, or too wide on both tacks, then the calculated true wind angle will also be too narrow or too wide (Diagram 3.2). Even this would not matter much – except that the true wind angle is used with the compass course to calculate the sailing wind direction. If the angles are wider or narrower than they should be the sailing wind

DIAGRAM 3.2 Errors in the apparent wind angle mean that the true wind angle (TWA) is calculated too narrow. Although the boat tacks through 60°, 2 × TWA = 50°. This means that the calculated sailing wind is backed on starboard tack and veered on port tack from the actual value of 240°.

will not meet in the middle – it will be different from tack-to-tack. Suddenly the defects in your calibration become all too clear.

The second approach accepts this, but then goes on to try to calibrate the sailing wind via the calibration of the apparent wind speed and angle. The way it works is that you tack the boat and see whether the sailing wind direction is different from one tack to the other. If it is, then your measurement of the apparent wind speed and angle must be wrong (assuming that the compass and boat speed are calibrated correctly) because the wind triangle is coming up with an impossible solution. By adjusting the calibrations of these two numbers you can change their values in the wind triangle calculation of the true wind angle. When they are adjusted correctly the true wind angle will give a sailing wind that stays the same as you tack. So you keep tacking the boat, adjusting the apparent wind calibrations and checking the sailing wind direction.

My feeling has always been that this is too difficult; it takes a lot of practice to be able to adjust one number, which only indirectly effects a second, to set up the second number precisely. That's why my preferred solution is to leave the apparent wind alone and not worry about the errors induced by the sail plan. After all, it is no less useful because it is a few degrees bigger or smaller, so why not leave the correction till you get to the true wind angle where it is important and it can be calibrated directly and simply? We will have more of this later when we come on to sailing wind calibration.

In the meantime, we are left with the relatively simple job of calibrating the apparent wind speed and angle. The apparent wind speed could hardly be easier. We have already mentioned that there is no possible way of comparing it to any other measurement of the apparent wind speed on the boat, so you calibrate it to the manufacturers' value and leave it at that. For those with the resources and the inclination, the only alternative is to take the unit off the boat and put it in a wind tunnel to check its measurement against a known wind speed. Some of the manufacturers provide this service for their units at quite a reasonable cost.

Unfortunately, the matter is not quite so simple for the apparent wind angle, and this is because of another phenomenon that we will be hearing a lot more of – wind sheer and gradient. This is due to a law of physics; moving fluids slow down when they come into contact with the friction of a solid. In this case the moving fluid is the air, flowing around the earth, and the wind is increasingly slowed as it nears the ground. This is termed wind velocity gradient – the different velocity of wind at different heights.

The slowing of the air is accompanied by a change in direction, in the Northern Hemisphere the increasing friction backs the wind, i.e. rotates its direction anti-clockwise. In the Southern Hemisphere the friction veers the wind – but in all further references to this effect we will assume we are north of the Equator. This effect is called wind sheer (Diagram 3.3). None of which would be a problem to the instruments if the effect was constant, but of course it is not.

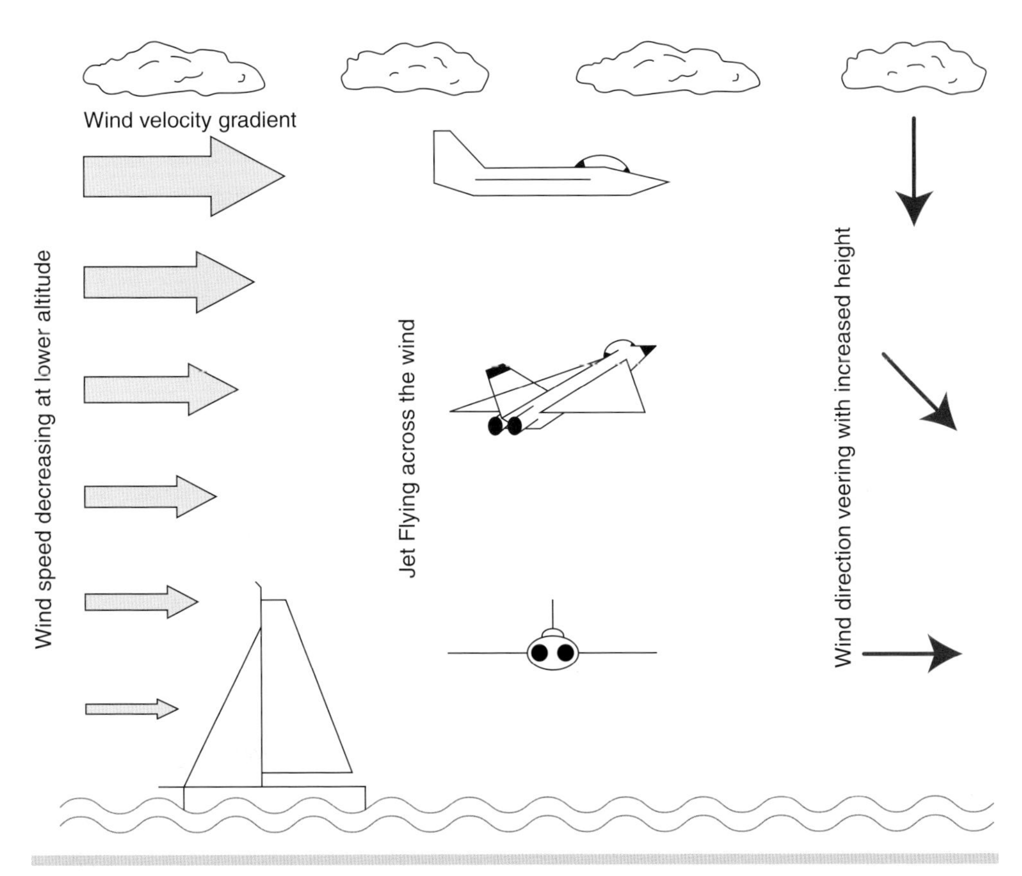

DIAGRAM 3.3 Wind Sheer. In the Northern Hemisphere wind direction veers with increased height and south of the equator it backs.

So how does this affect the apparent wind angle? When we calibrate it we are trying to get the unit to read zero when it is pointing up the centre-line of the boat, so that whether we are on starboard tack or port tack, if the angle we are sailing to the wind is the same, then the apparent wind angle reads the same. Wind sheer confuses this considerably – most obviously because the apparent wind angle will not be the same all the way down the mast. The sails will have to be set to some kind of average wind that is over the whole sail plan; the angle at the top of the mast need not bear any relation to this average sailing angle. Because the wind always backs as it gets nearer the ground what we will see is wider apparent wind angles on starboard than on port, when to all intents and purposes we are sailing at the same apparent wind angle as far as everything else on the boat is concerned.

We can only deal with this by calibrating the apparent wind angle when there is no, or at least very little, wind sheer. The idea is to leave it fixed at this calibration even when there is wind sheer. Although the numbers seem crazy, it is better to know that the instruments are right and that you are looking at a physical effect than to be constantly trying to change the instruments to match a fickle wind. Now we have a chicken and egg situation – we need to be able to tell whether or not there is wind sheer so we can calibrate the instruments – without having the instruments to tell us.

The critical factor in the amount of wind sheer and gradient is the vertical mixing of the layers of air. If the air is turbulent, then fast moving air from above will be mixed with the ground level wind and there will be little wind sheer or gradient. However, if the air is not turbulent, then there is little vertical mixing and nothing to prevent the air closer to the ground from being slowed and changing direction. So what causes this mixing? The answer is usually some kind of thermal effect; the air near the ground is heated by the land (heated in turn by the sun) and so it rises. When it rises it cools again and drops back down to lower levels. This mechanism mixes up the air. Another method is mechanical turbulence, which occurs as the air flows across hilly or rough ground. Trees, buildings or mountains, or even rough seas, will all start the air rotating and mixing.

We need to know the physical signs of these types of mixing so we can pick a good day to calibrate. In the case of thermal mixing it is relatively easy; cumulus clouds are created by the rising hot air, giving us some unmistakeable evidence. Mechanical mixing is a little more difficult to spot, but any weather system wind that is accompanied by clouds or a frontal system will usually be well mixed. This is because it has usually been travelling for some time across sea or land. The days you want to avoid are the high pressure days when there are either uniform clouds or a clear sky, often accompanied by light winds.

There are other signs you can use associated with the boat. The wind sheer and gradient will make the boat feel different on port tack to starboard. Typically you will be able to sail more quickly on starboard tack with the sails set with looser top leeches. On port tack, speed will be much more difficult to get and the sails will need tight upper leeches (Diagram 3.4). The reason is the change in direction of the wind with height. The wind is backing as it gets closer to the ground on starboard, so the wind angle is wider at the top than it is low down – hence loose upper leeches and plenty of power; whereas on port the top of the sail will be at a wind angle that is actually too tight – hence stalled and low power sails with tight leeches. Look out for these signs and do not calibrate when you see them. Check out the 'Don't Panic' section too to see how you can make use of this information.

When you have chosen your day to calibrate the routine is simple enough. First, sail the boat head-to-wind, having the mainsail up will make it easier to tell when this is the case. The apparent wind angle should read zero, if it does not then reset the calibrations so it does. Next, set the boat up on starboard tack so that it feels comfortable. Make a note

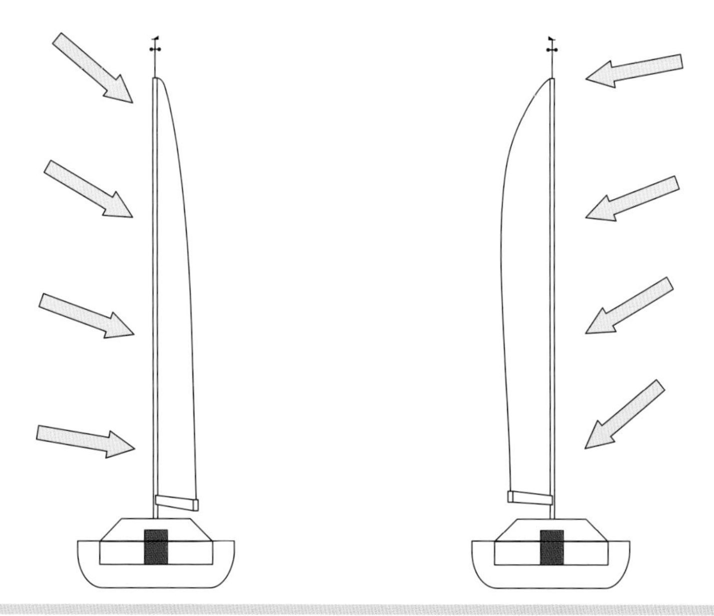

DIAGRAM 3.4 The effect of wind sheer on sail trim; starboard tack with open top leeches and port tack with tighter top leeches – and vice versa in the Southern Hemisphere.

DIAGRAM 3.5 Apparent wind angle calibration.

of the trim of the sails and the boat speed that you are sailing at. Watch the wind angle for a while and take readings so that you can average them out to a single apparent wind angle number for that tack. Then tack the boat over onto port and set her up in exactly the same way – same trim and the same speed. The idea is that if these things are equal then you are sailing at the same apparent wind angle; in which case the apparent wind angle reading should be the same as well. Watch it for a while and average the numbers. If you have done the head-to-wind test properly then they should be close to the same. But the tack-to-tack test is more accurate, so if the calibration needs adjusting a little bit go ahead and do it (Diagram 3.5). Keep repeating the exercise until you are confident that the apparent wind angle readings are the same on both tacks. Something to watch out for is the sea state, if it is different on the two tacks then even though you are sailing at the same speed you will not be at the same wind angle – take care.

CALIBRATION OF THE DEPTH

The depth is one of the more straightforward calibrations. This is partly because it is not connected to any of the other measurements through the wind triangle, and partly because there is a fixed measurement to calibrate it to – the depth of the water. Having said that the depth can be a rather fuzzy distance if it is a muddy bottom, so try and find a solid seabed to do it on. The best method is to use a lead-line when the boat is in the dock. Measure the depth and then set the datum up so that the depth sounder reading matches it. Depending on whether you prefer to read actual depth or the water under the keel, you will need to include the distance from the depth sounder itself to either the bottom of the keel or the waterline. You will need to know this distance, or measure it while the boat is out of the water. One thing to watch out for is that depth sounders sometimes have trouble getting clear readings in crowded marinas. If it is reading erratically in any way, wait till you get a calm day and do it whilst you are stopped somewhere outside.

CALIBRATION OF THE HEEL ANGLE GAUGE

Although heel angle does not have a direct impact on the wind triangle calculation, it is used in the calculation of leeway (see Leeway Section below) and leeway is sometimes then used in the calculation of the sailing wind direction. Even if you are not sure whether this is the case, if your system has heel angle it would pay to calibrate it at the same time as the boat speed, compass and apparent wind, i,e. before calibrating the true and sailing wind.

The calibration is straightforward. On a calm day set the boat up with slack warps in the dock and put all the gear in its normal sailing position – including boom and spinnaker pole on the centre-line. Whoever stays on-board should also stand on the centre-line while they read the heel meter. Under these conditions the heel angle should read zero; if it does not then adjust it till it does either with a software calibration or, if one is not provided, by moving the unit itself.

CALIBRATION OF LEEWAY

Leeway is another function, like the true and sailing wind, that is calculated from measured data. The formula used may vary from one instrument system to another. It will almost certainly depend on boat speed and heel and should be quoted somewhere in the manual. Let us assume that it is as follows:

$$L=(K \times H)/(Bs \times Bs)$$

Where

L = Leeway angle
K = Leeway coefficient
H = Heel
Bs = Boat speed

The heel angle and the boat speed can be measured directly by the instrument system; the calibration number we need to find is actually the leeway coefficient. There is an easy way and a hard way to do this. The easy way is to ask the yacht's designer the value of the leeway coefficient; the hard way is to try and measure it. Somewhere in the middle lies a short cut which involves taking a guess at it, and then watching the leeway angle calculated by the instruments while you are sailing and checking to see how it matches the design predicted figures. This may not sound very thorough, but the measurement of the leeway coefficient is so difficult that it's probably as good a way as any.

If you cannot get any figures from the designer you can always try and measure the leeway coefficient. The idea being to measure the leeway angle for a particular heel angle and boat speed. Knowing these three numbers you can rearrange the above formula to give you the leeway coefficient, for example:

$$K = (L \times Bs \times Bs) / H$$

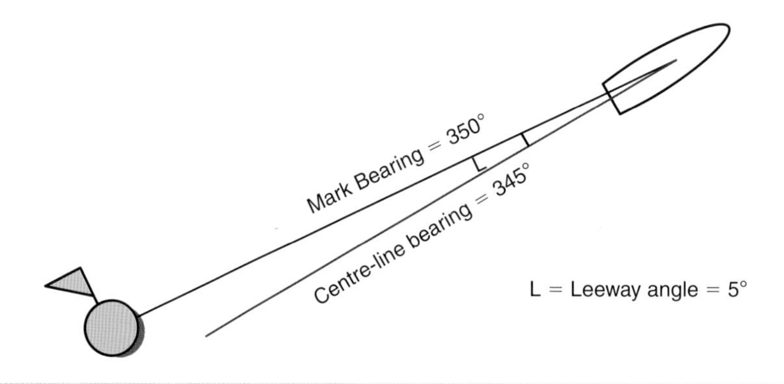

DIAGRAM 3.6 Calibration of Leeway.

The problem arises when you actually try to measure the leeway angle. This is how it is supposed to work (Diagram 3.6). You pick a day with around 10 to 15 knots of wind, steady in direction, with reasonably flat water. Sail upwind on a steady bearing, recording the boat speed and heel angle as you go. Throw a marker out of the back of the boat. Stand near the mast and using a hand-bearing compass, take a bearing down the centre-line towards the stern. From the same position take a bearing of the marker you dropped overboard. The difference between the two is the leeway angle.

This is combined with the boat speed and heel angle that you have been recording to calculate the coefficient. It goes without saying that the flatter the water and the steadier the breeze the easier this is, and even then it is not easy. Personally, I usually resort to taking a stab at the coefficient, and then adjusting it so that the leeway angle that the instruments calculate is about 3 degrees when we are sailing upwind in about 10 knots true wind speed and flat water.

CALIBRATION OF THE TRUE OR SAILING WIND

The calibration of the true and sailing wind is something that we have already discussed within the section on apparent wind calibration. I would certainly recommend that you read that before going any further here. Another section that you should look at is the one on the wind triangle. Briefly, what we already know is that the true wind speed and angle, and the sailing wind direction, are all calculated using the wind triangle and values of the boat speed, compass, apparent wind speed and angle that come from the sensors. Once you have calibrated these four sensors your instrument system will work perfectly adequately in every respect except two.

1. The true wind speed will read differently when you are sailing upwind compared to downwind.
2. The sailing wind direction – which is the main tactical tool that the instruments provide you with – will read differently from one tack to the other; and from one sailing angle to another.

The reason for this is that the masthead unit is prevented from measuring the apparent wind speed and angle accurately. A couple of factors combine to achieve this, the principle one being the deflection of the wind by the sail plan, but the twisting of the masthead unit by the mast is also a factor. So much we already know. We have decided to leave these errors in the measurement of the apparent wind speed and angle because they are invisible in these numbers. However, when the wind triangle calculates the true wind speed and angle we know the errors will reappear. Let's take the case of the true wind speed first.

We consistently find that the true wind speed reads higher when you are sailing downwind than when you are sailing upwind (Diagram 3.7). This is presumably because the airflow is accelerated past the masthead unit by the action of the sail to a greater extent downwind compared to upwind. Whatever the reason the effect is consistent, to correct it you need to take about 15 per cent off the values of the true wind speed that you see downwind. Some systems, instruments and computers, allow you to do this in a table so that the correction is always made for you. If your instrument system does not have the facility you will have to do the correction yourself each time the boat is

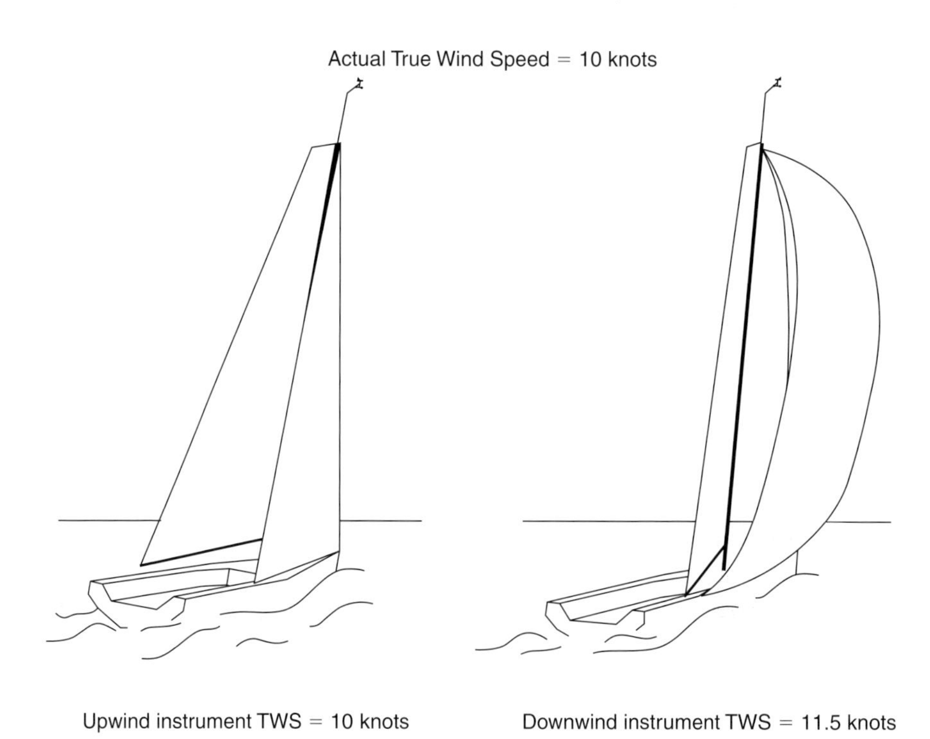

DIAGRAM 3.7 The Masthead Acceleration Affect when Sailing Downwind.

about to turn a corner. Note that this error will also exist in the apparent wind, but because the apparent wind is so different upwind compared to downwind it is almost impossible to spot the effect. Nevertheless, if you want an accurate downwind apparent wind speed you should also take 15 per cent off the value on the dial.

The true wind angle is not quite so straightforward. The reason is that the combination of errors that are present at the top of the mast cannot be relied upon to produce a consistent total error. It is dependent on the amount the mast twists, the exact position of the masthead unit – even the sails you put up will have an influence. On some boats the errors will combine to make the apparent wind angle smaller than it should be, on others it will be bigger. Whichever way it goes, the error will be carried through into the calculation of the true wind angle. Not that you will tend to notice it here – whether or not the true wind angle is accurate to within five or ten degrees will not be visible when you are using it for trimming or steering. It is when the instruments use it to calculate the sailing wind direction that you will notice the problem. The sailing wind is worked out by adding the true wind angle to the heading if you are on starboard tack, and subtracting it if you are on port. Obviously if the true wind angle is wider or narrower than it should be then the sailing wind direction will calculate to a different number on each tack (Diagram 3.8).

a. If the true wind angle is too narrow then the sailing wind direction will be backed on starboard tack compared to port.
b. If the true wind angle is too wide then the sailing wind direction will be veered on starboard tack compared to port.

A secondary problem exists in that the amount of error in the true wind angle, be it wider or narrower, is not the same upwind to downwind. If you bear away from sailing upwind to reaching, on the same tack, you will find that the sailing wind direction changes even though the wind is steady. We can conclude, firstly, that the true wind angle requires a correction at every point of sailing to make it read an accurate value. We need to develop a table of these corrections for sailing upwind, reaching and downwind, at a band of wind speeds. Secondly, the way to develop the table is to use the errors we see in the sailing wind direction as we manoeuvre the boat to tell us the corrections we need to make to the true wind angle.

We start by sailing upwind, preferably on a day when the breeze is not too shifty, making a note of the true wind speed and angle, and the sailing wind direction. The boat is then tacked, and after the instruments have settled (see the section on Damping) the new

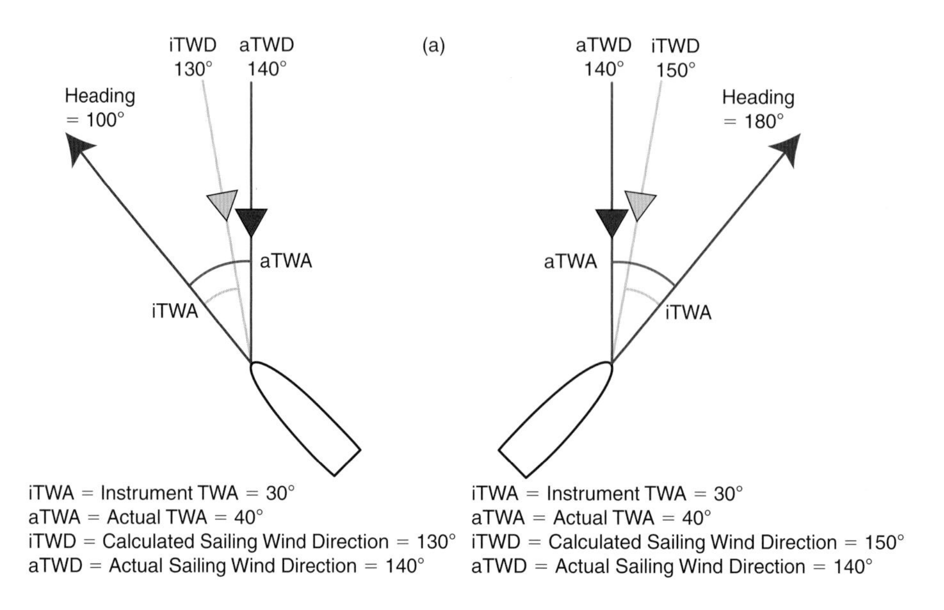

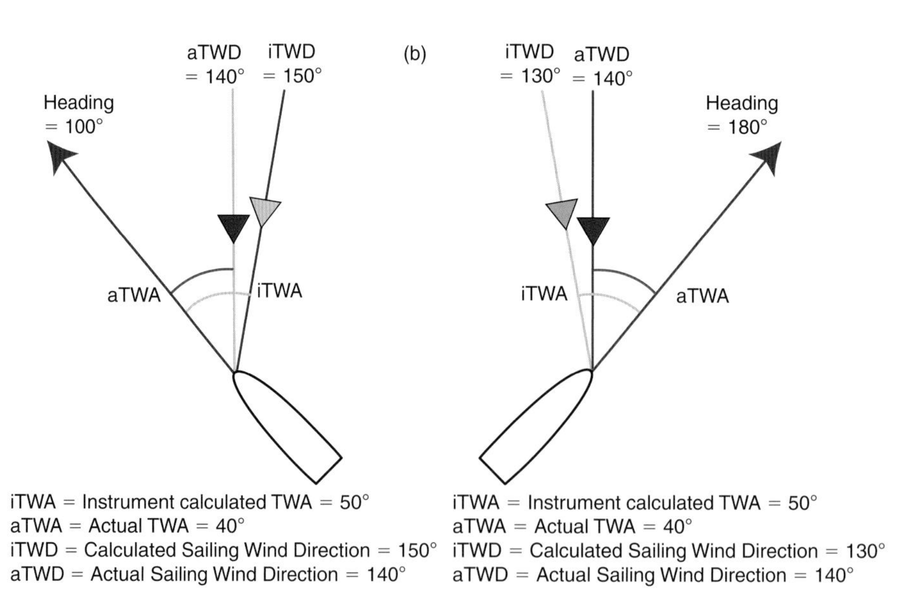

DIAGRAM 3.8 Calibration of the Sailing Wind Direction Upwind.

sailing wind direction is recorded. Check that the true wind angle and speed are about the same as they were on the other tack. Then tack back again and see if the sailing wind goes back to where it was before. You want to be sure that the wind has not shifted while you were tacking and any difference (or lack of it) in the sailing wind is genuine instrument error and not a wind shift.

Once you are satisfied, work out the correction to the true wind angle required. If the sailing wind direction on port tack is veered compared to starboard, i.e. 150 on port compared to 130 on starboard, then the true wind angle is too narrow (it is too wide if the sailing wind direction on starboard is veered compared to that on port). The amount of the correction is found by dividing the difference between the two sailing winds by two, and in the case where the true wind angle is too narrow, adding it on. So in this case we would add:

$$(150 - 130) \div 2 = 10 \text{ degrees}$$

to the true wind angle to correct it.

Once you are confident of the correction required upwind, try sailing along close-hauled and then bearing away to a close reach, perhaps a 60 degree true wind angle. Again you should watch the sailing wind direction for any changes and repeat the exercise several times to be sure that the changes are due to the instruments and not the wind. Assuming that you are on port tack, if, as you bear away, the sailing wind direction veers, then the true wind angle at 60 degrees is too narrow compared to the true wind angle upwind. And this means that you must add as a correction the difference between the two sailing wind directions.

Let's take an example (Diagram 3.9) – sailing along upwind on port tack and the sailing wind direction reads 200, you bear away to 60 degrees true wind angle and the sailing wind direction changes to 210. Then we know that we must add the whole of the difference between the two sailing winds, which is 10 degrees, as a correction to the true wind angle at 60 degrees.

Another example (Diagram 3.10) – we are now on starboard tack sailing at 120 degrees true wind angle. The wind speed is steady and the sailing wind direction is 300 after it has been corrected upwind. We bear away to 160 degrees true wind angle and the sailing wind alters to 315. What is the correction required for the 160 sailing angle?

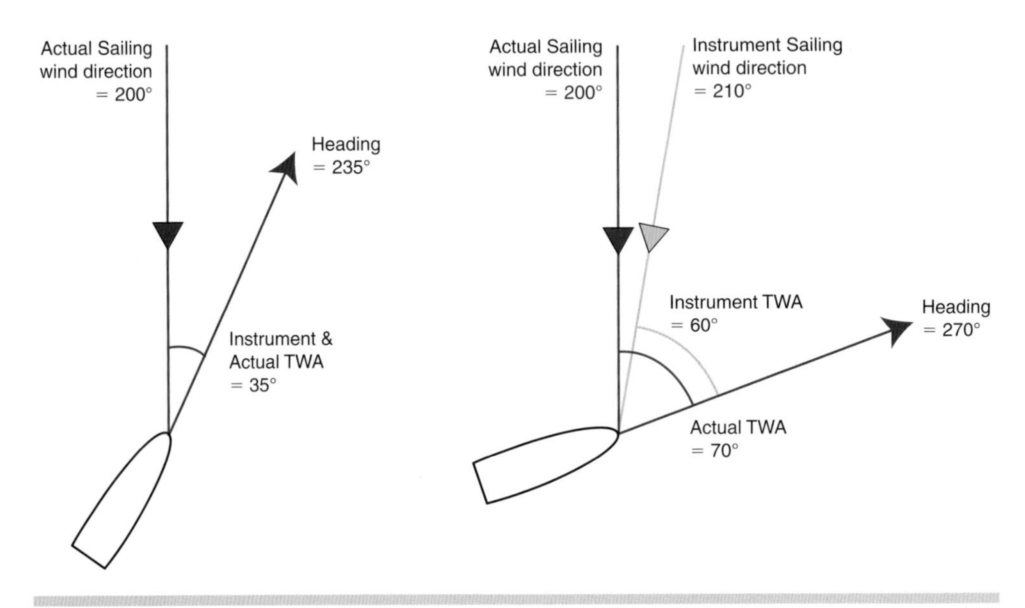

DIAGRAM 3.9 Calibration of the Sailing Wind Direction on a Reach.

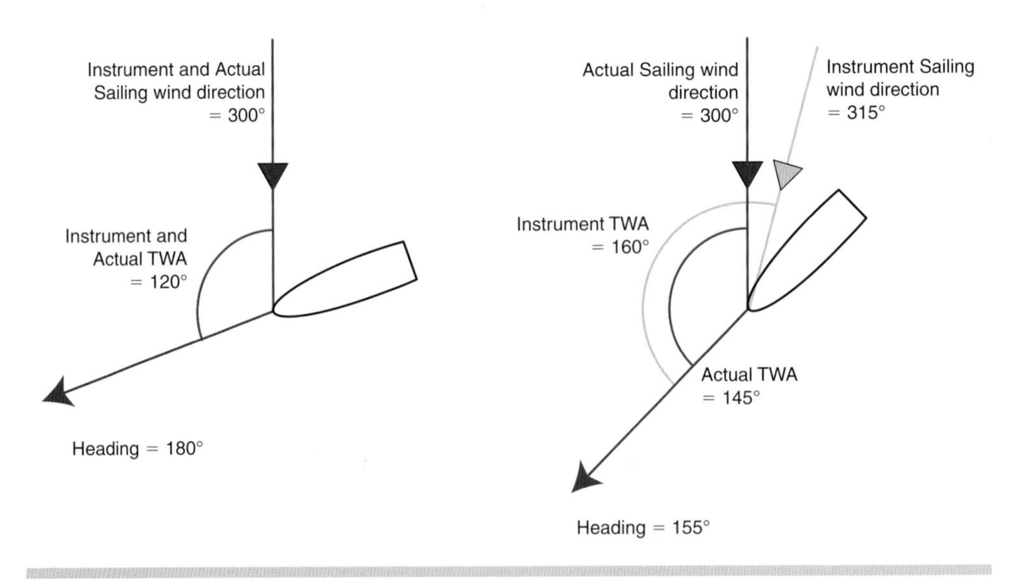

DIAGRAM 3.10 Calibration of the Sailing Wind Direction Downwind.

The correction is 15 degrees; the difference between the two sailing wind directions. Now we just need to work out whether it should be added or taken away from the 160 true wind angle. Because we are on starboard and the sailing wind direction veers, this means that the true wind angle is being calculated too wide. So we should subtract the 15 degrees to get the right answer. Another way of thinking about it is to work out what you need to do to the true wind angle to change 315 to 300, and the answer must be, for starboard tack, to subtract 15 degrees. Eventually, you must slowly and repeatedly manoeuvre the boat through the full range of sailing angles, carefully calculating the necessary corrections to keep the sailing wind direction the same.

Once you have the table of corrections you need to apply them to the true wind angle. Some instrument systems have built in tables that allow you to enter the corrections into them. The instruments then automatically apply the corrections for you, depending on the wind speed and angle you are sailing at. Some on-board computer systems will do the same job, and it would not be too difficult a task for the computer literate to program a laptop to do it. But failing all that you are stuck with pencil and paper – perhaps a correction table, laminated and stuck to the deck somewhere.

It's still really useful; take the situation where you are coming into a port round leeward mark right behind another boat that you are catching fast (Diagram 3.11). You need to know whether you want to start the next beat on port or starboard so you can choose which side to overtake him on. Not unnaturally you look at your sailing wind direction to see which tack is favoured and it says the wind direction is 145 degrees. From your wind readings up the last beat you know that this is a big port tack-lift. In which case you would be happy to go round outside him at the mark, so long as you can get clear air rather than push for an inside overlap.

However, your correction table tells you that you should subtract 20 degrees from the true wind angle to get the same sailing wind direction that you had upwind. So you subtract 20 from the true wind angle on port and this veers the sailing wind direction round to 165 degrees. It puts rather a different complexion on things now that you have a big starboard tack-lift. You promptly slow the boat down and round the mark neatly behind the other boat. Now you can tack immediately after the mark and get on that lifted starboard tack. Without your correction table you would have gone round the outside of the other boat, then watched as the sailing wind direction changed to its upwind value and realised you were on the wrong tack. But the other boat is pinning you on port and until you can clear him you are stuck in the header – no way to start a new beat.

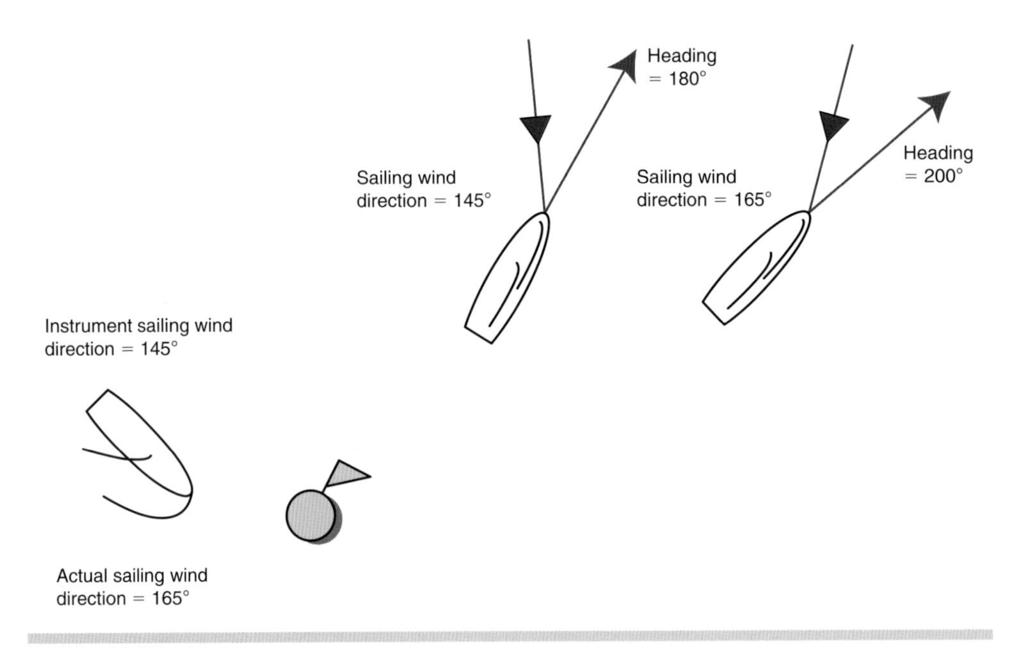

DIAGRAM 3.11 Planning your tack away from the mark using an inaccurate wind direction will put you on the wrong shift.

One final comment and this is the bad news. There is absolutely no guarantee that once you have worked out your true wind angle correction table that it will work for more than a few minutes. What?! I hear you cry, understandably aghast that so much work can so easily be destroyed. Well yes, it can. The conditions of wind sheer and gradient that you do this calibration in seem to have an effect on the results. Quite why this is I can only guess, but you can imagine how a much lighter breeze at water level might be deflected more than a stronger one – and so you need more or less correction for the same wind speed value at the masthead.

The only consolation is that, given typical conditions at a particular venue, the corrections seem to remain about the same, and when changes do occur they seem to be in the size of the correction rather than its direction. So once you have got your table for all the sailing angles it should work most days. Doing a few tacks before the start to check it out is definitely recommended. If the weather is unusual or you are sailing at a different venue, i.e. the Mediterranean rather than the Solent, then expect some very different numbers.

Wind sheer has a more obvious effect on the accuracy of the sailing wind direction. It rotates the numbers by however many degrees of sheer there is. So if you have 20 degrees of wind sheer, which you can see from your apparent wind angle readings, then the wind you are sailing in is up to 20 degrees different to the sailing wind direction you are reading on the dial. A good way to check this is to work out what the wind should be from your headings on both tacks. If you are sailing at 200 on port tack and 270 on starboard tack then the wind should be blowing from 235 – half-way between the two headings. If the sailing wind dial reads 245 then you have 10 degrees of wind sheer. Another way of checking before the start is to go head-to-wind with the boom flapping on the centre-line and see if the compass heading is the same as the sailing wind direction.

The main reason why you need to watch out for wind sheer errors is next leg calculations, which we will discuss in Chapter 6. But it is fairly evident that if you are using a wind direction that is 20 degrees different from the one you are sailing in it will throw out any calculations you might do for the next leg. At least it is an easy problem to deal with, just subtract the right number of degrees until the wind sheer goes away.

DON'T PANIC!!

It's a long time ago, but the most extreme and difficult conditions I ever faced were in Kiel Week way back in 1989. It was the British trials for the Admiral's Cup team and the first real racing for a very carefully set-up instrument and computer system aboard the IOR fifty footer, Jamarella. It was half way through one of the inshore races, and according to the instruments we were sailing straight into the wind with an apparent wind angle of zero degrees and a breeze of 10 knots. Meanwhile the sails were full and the water was a glassy calm. Confused?

Don't be – although this is a very extreme example, such apparent instrument anomalies are relatively commonplace. The crew's response is just as common, 'There's something wrong with the instruments.' It's a phrase that has a special place in all my worst nightmares. But in this instance there was nothing wrong with the equipment. It was still telling us something useful but the message needed more interpretation. Producing good information when you cannot read the number directly off the dial is one of the key skills of the navigator. So what was going on that day in Kiel?

It was our friends from the apparent wind calibration section – wind sheer and wind gradient. There was little or no mixing of the wind aloft and down on the water so there were big differences in wind speed. It is perfectly possible that the wind 80 ft up is travelling at 10 knots, whilst at zero feet it is stationary – and that's what the instruments were telling us. The masthead unit can only measure the wind speed where it is.

This information can be of use to the trimmers, they need to set the sails flat at the bottom for very light airs and full at the top for 10 knots. But we also need to take into account the wind sheer. We already know that as the wind slows, it backs, so that the wind at the masthead is considerably more veered than the wind at the water. This was why we were able to fill the sails on Jamarella even though the wind at the masthead was coming from dead ahead. The wind halfway down the mast was blowing from further to the left (we were on port tack) and filling the sails. Again the instruments were telling us all they knew, and the information was certainly of interest to the trimmers. It just needed interpreting correctly. So the next time the trimmer says 'There's something wrong with the instruments.' Don't panic! Not only are they quite possibly wrong, but they are the people who need the information most.

DAMPING – HIGH OR LOW?

Most instrument systems provide some facility for damping the data that they produce. The damping (or filtering, to use the more technical name) does just what the name suggests – controls the speed of response of the numbers you see on the dial to changes in the raw data. The maths of damping can be done in lots of different ways, but the simplest is to average over a variable period of time the data coming from the sensors. The shorter the period of time the quicker the values on the dial will respond to changes. This can be advantageous for seeing changes quickly, but you may well find that the numbers jump around so much that it is impossible to tell what they mean. In this case you need a longer damping period to average out all the small, quick changes so that you get a clearer view of the overall picture. Usually there is a happy medium between quick response and smooth changes, but it will be different for different conditions. In big breezes and waves the boat, and therefore the instruments, will be jumping around a lot more and so the numbers will need more damping. In light airs and flat water you can bring the damping right down so that you can pick up the changes in those zephyrs really fast.

If your system only has damping for the sensor values, i.e. the boat speed, compass, apparent wind speed and angle; then to change the damping of one of the calculated values you will have to change the damping of all the numbers in the calculation. So for true wind angle, it would be boat speed and apparent wind speed and angle; for sailing wind direction you will need to change the compass as well. Of course this may mean that these numbers then jump around too much, particularly the boat speed which the helmsman is trying to sail to – in which case you will have to compromise. The time it takes for the sailing wind direction to settle on its new value after a tack is one of the biggest bugbears for the tactician. The calculation of the sailing wind direction involves more components than any other number, and so the damping of all those values accumulates to make it slow to settle. It can easily take half a minute after a tack is completed for the sailing wind direction to find its new value. It is really important to remember this, reading it too soon will lead to problems whether you are calibrating, wind tracking before the start, or on the race course.

Some Instrument Techniques

START LINES AND WIND SHIFTS

The start of the race will be the first test for your instrument system. You will have two main jobs to do; the first will be to work out which end of the line has the advantage, and secondly, which is the best tack out of the start line. For both of these tasks you will need to use the sailing wind direction. We will leave aside such concerns as general strategy for the course, be it tidal or wind, and any impact this might have on the end of the line or the first tack. The instruments cannot really be expected to help you with this, although we will see later that a computer system can.

Choosing which end of the line to start from means choosing the one closest to the wind. The easiest way to work it out is to take a bearing along the line. Then add 90 degrees if you took the bearing from the starboard end, or subtract 90 degrees if the bearing is from the port end (Diagram 4.1). This gives you a value that I call the neutral line wind direction. It is the sailing wind direction that is completely square to the line, i.e. there is no advantage to starting at one end or the other. If the wind veers from this then the starboard end will be favoured, and if it backs the port end is favoured. Once you know the neutral wind direction all it takes is a glance at the sailing wind direction to tell you the favoured end.

Now you need to track the sailing wind direction down to start time. Using this information you can then pick the end of the line in time to get there for the gun.

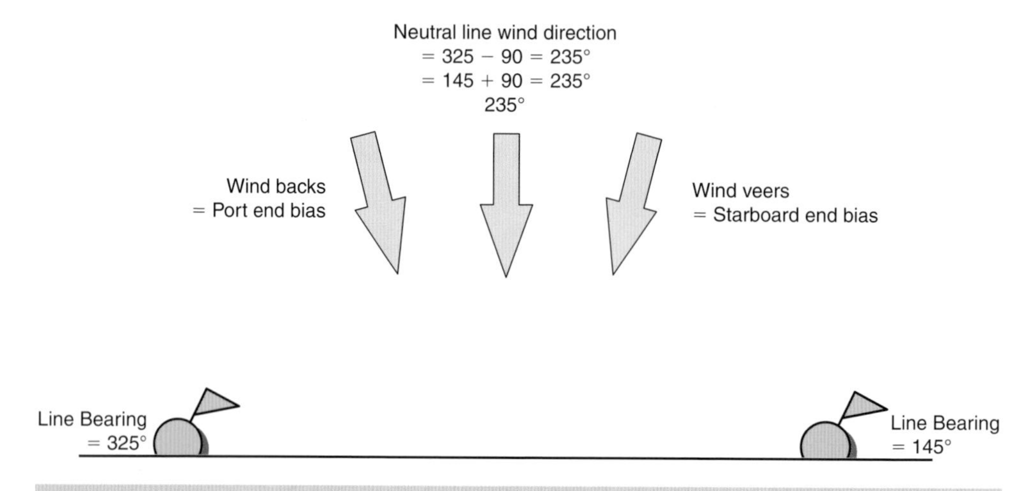

DIAGRAM 4.1 Calculating the Neutral Line Wind Direction.

Remember that if the wind is oscillating you may need to anticipate the shift that you will be on at start time. Imagine a situation where at nine minutes to go the port end is biased; but by six minutes to go the wind has swung and the starboard end now gets the nod. With three minutes to go it is back to the port end – but if you made the decision to head that way you might well find that at the start gun the starboard end is favoured.

Tracking these shifts down to start time is more or less just a question of writing down the time and the number on the display. But we must keep in mind all we have said in the previous sections about calibration and damping. You must check the calibration of the sailing wind before you start taking wind readings. Calibration errors are often particularly severe when the boat is tacking from reach to reach – which tends to happen quite a lot before the start. Don't be fooled into thinking that there are huge shifts around when there are not. Damping is another problem when the boat is being thrown around in the starting area; particularly when it is combined with lots of dirty wind from all the sails and confused seas. In fact, when it comes down to it you will be hard pushed to get a decent reading when you get into the final approach to the line. So it is really important that you have a clear idea of what wind directions you are expecting to see once you clear the line.

The start is a time when you must try to divorce yourself from everyone else's immediate concern – which is getting a good start – and look ahead to the first couple of minutes of the beat. Your first job is to work out what the wind is doing as you come off the line. You have to know whether the instruments are settled on the number they are showing or just spinning past it as the damping tries to cope with some radical manoeuvre the boat has just executed. In short, you need to watch it all the time. It is never easy to ignore the excitement of a start, but you will look pretty average when, as soon as you are off the line, the tactician turns round and says, 'Are we up or down?' and you do not know.

One complication you need not worry about is the effect of the tide on the start line wind. Sometimes an inexperienced race officer will set a badly biased line. The reason for this is that he is measuring the wind direction from a boat that is anchored to the seabed – and so it is the ground wind that he is recording. You are sailing in water that may be moving relative to the seabed and so your sailing wind will have a tide wind component (as we mentioned in the section on the wind triangle). This tide wind component can alter the wind you are sailing in quite dramatically (Diagram 4.2). If you ignore the tide and set the line to the ground wind you may well have a substantial

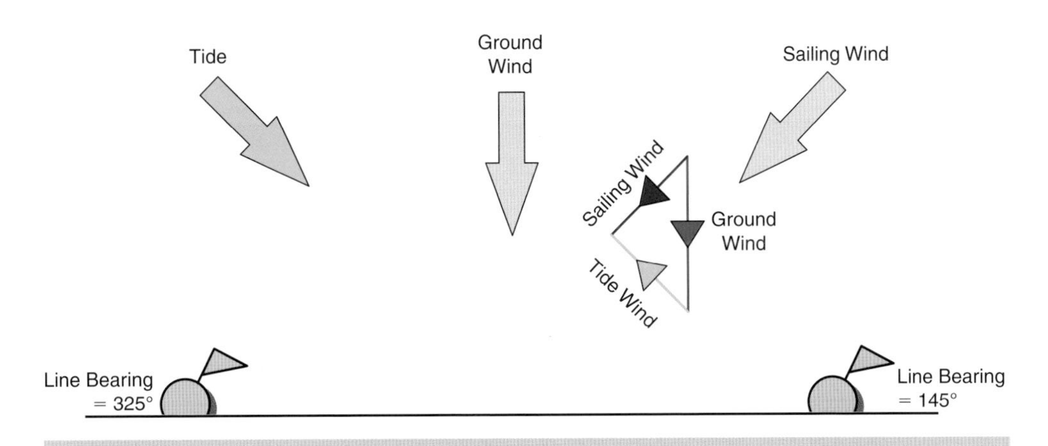

DIAGRAM 4.2 The line would be square to the ground wind, measured from the anchored committee boat, but has substantial starboard bias to the tidally altered sailing wind.

bias. The good news is that your on-board instruments (assuming that you are not also anchored) will read the sailing wind that includes the tidal component. It is only the race officer, who is anchored, that must account for it in his calculations.

One final point about wind shift tracking. We had mentioned earlier that changes in wind gradient and sheer may affect the calibration of the sailing wind direction. This poses a problem; how are you supposed to recognise calibration that has altered whilst you are racing? If, for instance, it started altering when you tacked in a manner that your calibration did not account for, how would you know? Would you not just assume that the wind was shifting as you tacked? For a while you might, but after three or four times you ought to be suspicious. Three or four tacks on dud information could cost you the race, so here is a check you can use. Keep a note of your compass headings on each tack as a back-up. These will also tell you if you are headed or lifted. As soon as you are worried that the sailing wind direction is playing up you can check what it is telling you against the heading.

Unfortunately there is a problem with this as well, since the true wind angle a yacht sails at is dependent on the wind strength. The stronger the wind the closer you can sail to the wind, until about fifteen or twenty knots when the angle does not get any narrower and may even widen as the wind increases towards thirty knots. Your compass will often tell you that you are headed or lifted when in fact the wind direction is the same, but the wind velocity has altered. This is known as a velocity header or lift, and is accentuated by what happens to your apparent wind when the sailing wind first changes. Diagram 4.3 shows the case of a velocity header. As the wind drops, the boat has sufficient momentum to keep its speed for a few seconds, which moves the apparent wind forward, lifts the jib and gives the impression that you have been headed. It is important that the helmsman does not bear away too hard when this happens because as the boat slows down to match the new wind speed the jib will stop backing, and you can gain ground to windward by holding course and letting the speed drop until the jib refills. You will have to bear away a little because of the new wider true wind angle for the lower windspeed.

Do not be fooled into tacking by the velocity header, and it is equally as important to be careful when checking your sailing wind direction against the compass heading so that you do not start worrying about the sailing wind direction unnecessarily. The sailing wind will ignore velocity headers and lifts, whereas the compass will not. So you really have to keep an eye on them both, each to check the other. In the next section we will see how we can use the heel angle to help with a similar problem with the true wind speed.

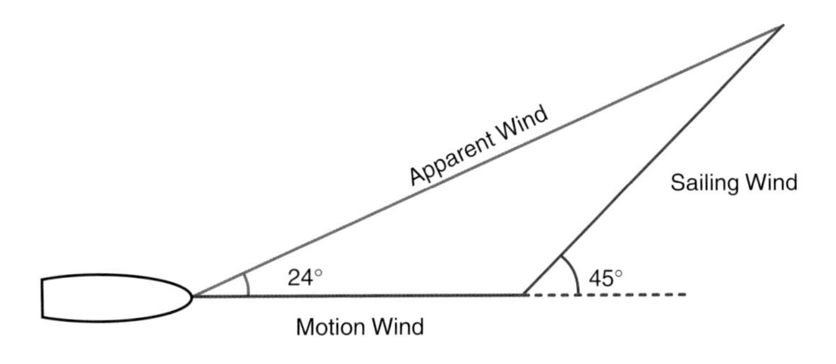

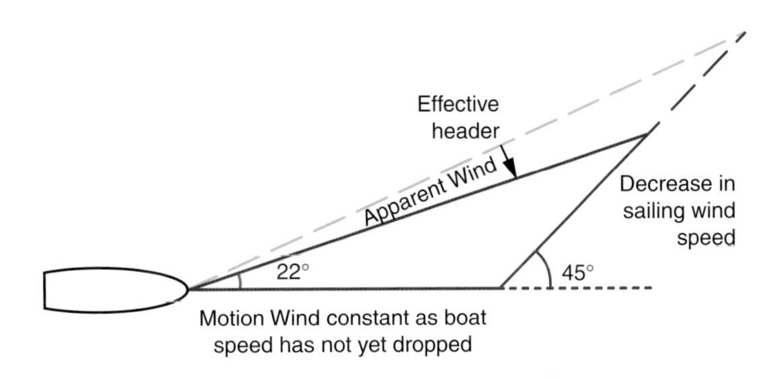

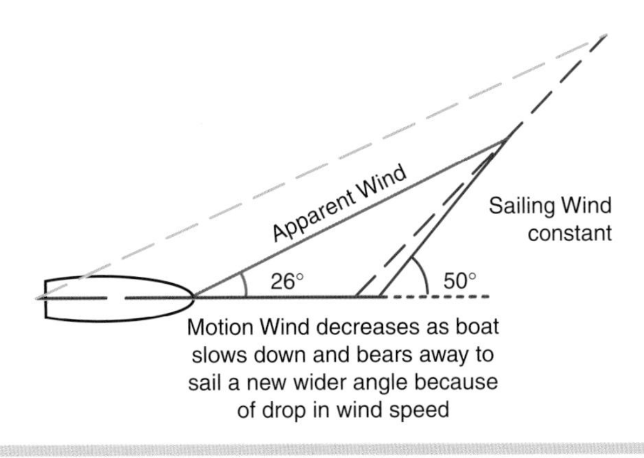

DIAGRAM 4.3 The Velocity Header.

HEEL ANGLE OR WIND SPEED?

In the Don't Panic section we looked at an example, albeit extreme, where the wind speed read ten knots at the top of the mast and the water was a glassy calm. The effect of wind sheer and gradient meant the instruments required careful interpretation. We looked at how to deal with the apparent wind angle, but not the wind speed, which you need for sail selection and target speeds. So what do you do when the wind is seriously mixed up about how windy it is? The answer is to use the heel angle.

At any sailing angle other than downwind, the heel can be an excellent measure of how much power there is in the wind. This is what you are using the wind speed to tell you so that you can match your sails to the wind available. Under average conditions of wind gradient it is a reasonably good guide. It will, however, always be limited by the fact that it is only measuring the wind speed in one place, and the wind is quite capable and equally likely to change at a height other than the masthead. In this case, the only way the instruments will pick up the change in the force available from the wind is via the heel angle (Diagram 4.4).

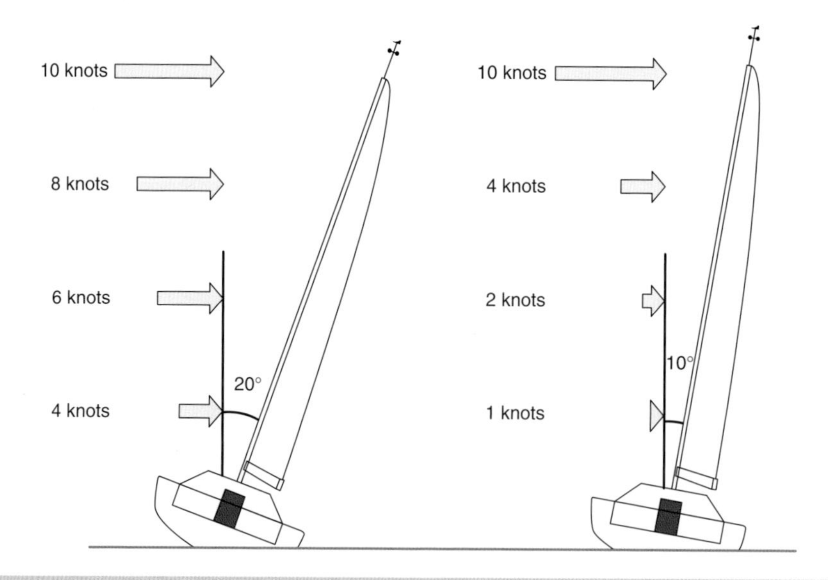

DIAGRAM 4.4 The Affect of Wind Gradient on Heel Angle.

Before I go any further I should put in a proviso about the heel angle not just being dependent on the wind speed; sail choice, trim, steering technique etc. will all affect it as well. But when you are sailing along in a straight line with all these things more or less constant, and suddenly the heel angle starts dropping – while the wind speed remains the same – it should be clear what is happening.

I have seen this type of thing most often in the Mediterranean which is prone to the light and fickle conditions where the technique is useful. It's a place where the wind sheer comes and goes, or bands of warmer air blow in which are less dense and therefore exert less force or pressure. This is when wind speed measurement becomes less than trustworthy. Keeping an eye on the heel angle can give you that vital first clue to what is going on. But do not expect the technique to work when you are crashing upwind in twenty knots, the heel angle will be jumping around far too much to be useful unless your instrument system has the facility to damp it. Fortunately, in these conditions the wind is usually steady and consistent at all heights, and the wind speed does the job it was intended for perfectly well.

ANTICIPATION

Anticipation is not so much an instrument technique as a state of mind for the user. Because of the greater emphasis on tactics in short course racing, you will spend less time working on strategy, and more supporting the tactician with the information he needs for his decisions. The key to doing the navigator's job well is to anticipate what the tactician is going to want to know next and start working it out before he asks for it. There are endless possible examples, and I have pointed out a couple below. But if there was one single piece of advice I would give to a big boat navigator on this subject, it is to do some small boat racing as a helmsman and decision maker. There is no faster way to get an insight into what the tactician needs to know. Then read all the books on yacht racing tactics as well as those on navigating.

The first example comes up all the time. You are on port tack heading towards the starboard tack layline and apparently on a collision course with two yachts on starboard tack. The tactician needs to know whether or not those boats are laying the mark before you get to them. He has to decide whether to duck behind them or tack underneath. If they are laying the mark comfortably, then a tack to leeward and slightly ahead will see you round in front. If they are not laying, it would be preferable for you to duck behind them and sail on to the layline. The navigator should see this coming way before it arrives, particularly in a tidal situation where it is much harder to judge the layline by eye.

Delving back into ancient history for the second example, I still remember the final inshore of the 1989 Admiral's Cup. A backing shift had come in on the first beat and we were sailing to the gybe mark on a tight reach, fourth of the fifty footers. The last of the leading three peeled round and gybed when we were about fifty yards from the mark. The tactician asked if we could carry on, meaning had the wind swung enough to make the next leg a run rather than a reach. If it was a run we could start it on either gybe, but preferably the one that was most advantaged by the present wind shift. I had been looking at the problem for a minute or so, it was certainly a run; the question was whether or not starboard put us on the best shift. By the time the question came I was able to answer yes. The gybe was cancelled and we squared away and carried on on starboard. A couple of minutes went by and the breeze lifted us, which downwind takes you away from the mark. So we gybed and, laying the mark on the paying tack, ran down to it and into second place.

The wind shift coming through when it did made us look particularly smart – but starting to look at the question before it was asked was the only way to have the answer ready in time. A good tactician will expect this sort of anticipation; it is not his job to be warning you of every possible situation that might arise. He is going to ask the question when he needs the answer, which is usually immediately. So keep your eyes on the race course, concentrate and anticipate.

It is a lot easier if you can program a computer or calculator to do all this for you, allowing you just to enter the next leg heading, wind direction and speed. Some instrument systems allow you to enter a course and from this they calculate the next leg data. Obviously the instrument system would have to have an internal polar table that matched your boat, and this is unlikely to be found in anything other than top of the range models. A simpler solution is to make sail selections from true wind angle and speed rather than apparent!

VELOCITY MADE GOOD TO THE COURSE: VMC

The concept of VMC, or 'Velocity Made Good in the Direction of the Course', is something you might have come across on your position fixer. It is a simple enough idea, being the net velocity that you are making towards the mark; in the same way that VMG is the net velocity that you are making towards the wind. It is calculated in a similar fashion, being the velocity that you are making across the ground multiplied by the cosine of the angle between your COG and the course to the mark. As in Diagram 6.2, if X is the angle between your COG and the bearing of the mark then:

$$\text{VMC} = \text{SOG} \times \text{cosine X}$$

All these values are available to a position fixer which has a waypoint memory, which is why some of them have started calculating the value for you. Whether or not you should use it is another question altogether.

VMC is a tactical and strategic tool that helps you follow the general rule, 'Shorten the distance between you and the mark as quickly as possible'. It is most valuable when you have no idea what the weather or the current is likely to do. If the geographical effects on the next leg are, to all intents and purposes, random, then your best chance is to keep taking the option that gets you closest to the mark. Whatever the weather throws at you, you have the least distance possible to sail – even if it is directly upwind! The point to bear in mind is that there is almost always a faster way to sail the leg than by just optimising the VMC. The way to achieve this is to predict correctly the affects up the leg and then position the boat to make the best use of them. We will look at one approach to this in the next section. But first there are a couple of points that we should

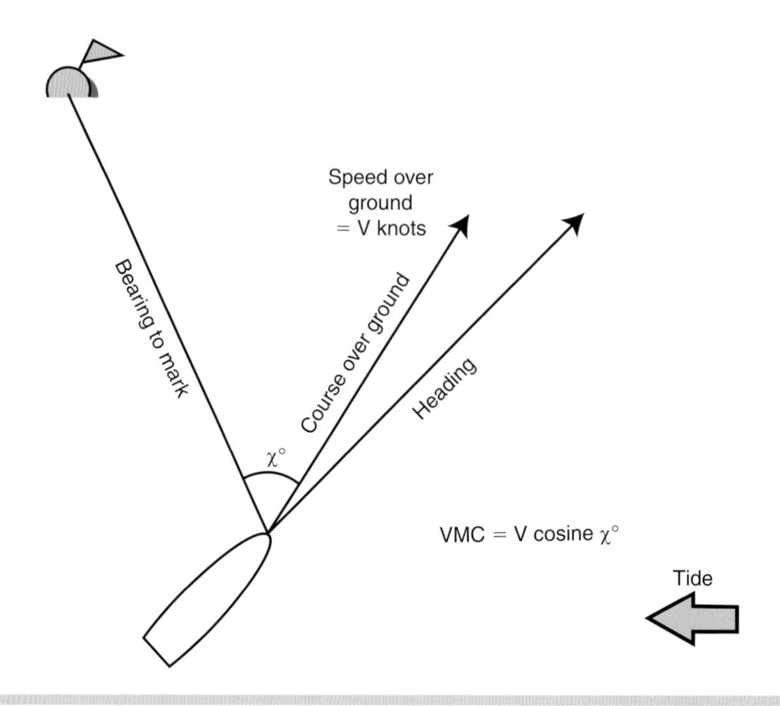

DIAGRAM 6.2 The Calculation of Velocity Made Good in the Direction of the Course.

make; optimising your VMC on a reaching leg may not mean sailing straight at the mark; secondly, optimum VMC upwind is not necessarily the same as your optimum VMG.

Let's take the reaching leg case first. It may sound rather far-fetched to say that you can close on the mark faster by not sailing on the direct line towards it, but it all depends on the shape of the polar curve (Diagram 6.3). Some polar tables have a pronounced bump that allow the extra speed gained by going faster and away from the mark to make up the extra distance sailed (VMG sailing works on exactly the same principle). You would be right in asking the obvious question, 'If you start by sailing off at an angle to the course, how do you eventually get to the mark?'. The answer is that as you sail down the leg, the VMC course and the course to the mark converge until they are the same. You end up sailing a loop to the mark. Whether this loop is faster in total than just sailing the rhumb line would depend on the exact detail of the polar table, and you will probably need a computer to work it out.

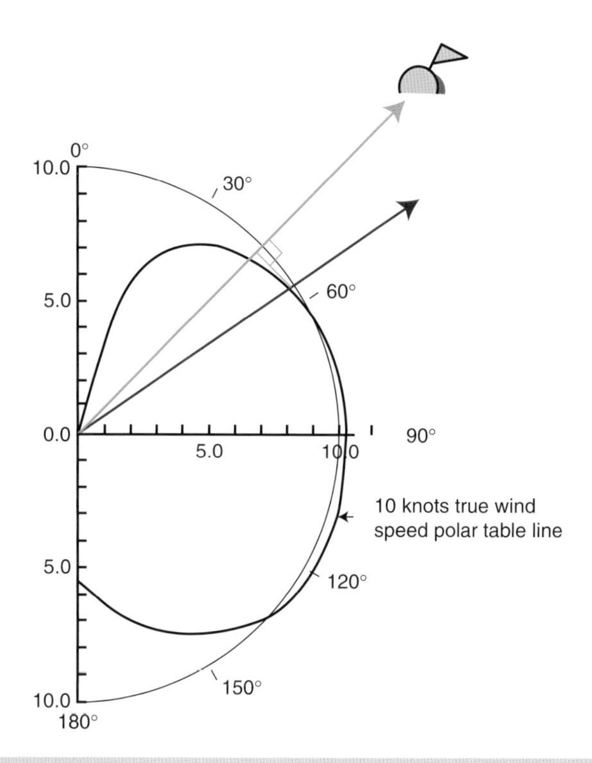

DIAGRAM 6.3 The Optimum VMC is not the Direct Course to the Mark.

Sailing to optimise VMC is really only going to work where the leg is a long one and you know the weather is going to change, but have no idea how. Let's imagine that you sail the optimum VMC and everybody else goes down the rhumb line. Half-way along the leg you are a couple of miles closer to the mark when the wind drops out. There are a few hours of calm before the race starts again with a new wind direction. That couple of miles is then converted into a lead – so long as you are not disadvantaged by the new wind direction compared to the opposition. However, given all that we have said about the approximate nature of polar tables, the wisdom of yachting off at a tangent to the rhumb line on a whim of the polar table does sound rather suspect...and I would have to confess that I have never done it.

This is a good moment to make a point about the use of polar tables in tactical situations. Despite all that we have said about the accuracy of polar tables, they do seem to work well in tactical applications. They are not too badly affected by the problems of wind sheer and gradient because they work from the shape of the polar curve more than from the exact speed predicted (Diagram 6.4). Wind sheer means that the boat speed achieved at any particular wind angle and speed is inconsistent; but within a wind speed range of four or five knots, the proportion of boat speed to angle represented by the shape of the curve is the same. Only extreme conditions of wind sheer

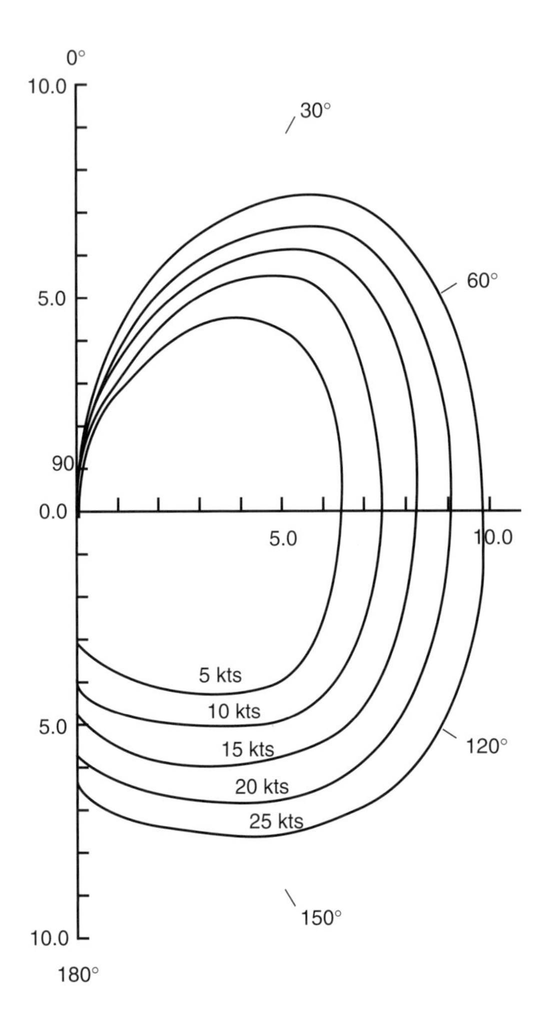

DIAGRAM 6.4 Because the shape of the polar table is similar for different wind speeds, wind gradient is not too great a problem for tactical applications of the polar table. One uses the shape of the curve, rather than the precise numbers from it that are required by performance uses of the table.

will produce a sufficiently big gap between the measured wind speed at the masthead and the real wind force available to make the polar table shape different – so tactical judgements from polar tables will work reasonably well, despite all the problems.

Going back to the reaching leg we had considered earlier, there is an occasion where sailing off-course is highly recommended. This is when the leg is just a little bit too tight to hold a spinnaker all the way down it, so that if you do hoist you will end up low of the mark. The alternative is a two-sail reach all the way down the leg. It is a situation that you come across quite a lot and it is much quicker to put the spinnaker up and hold it as long as you can (Diagram 6.5). You must judge the drop quite carefully so that you come in on a fast two-sail reach at the end. We can see why the longer course works if we look at the polar table. There is usually a concavity between areas of the curve where

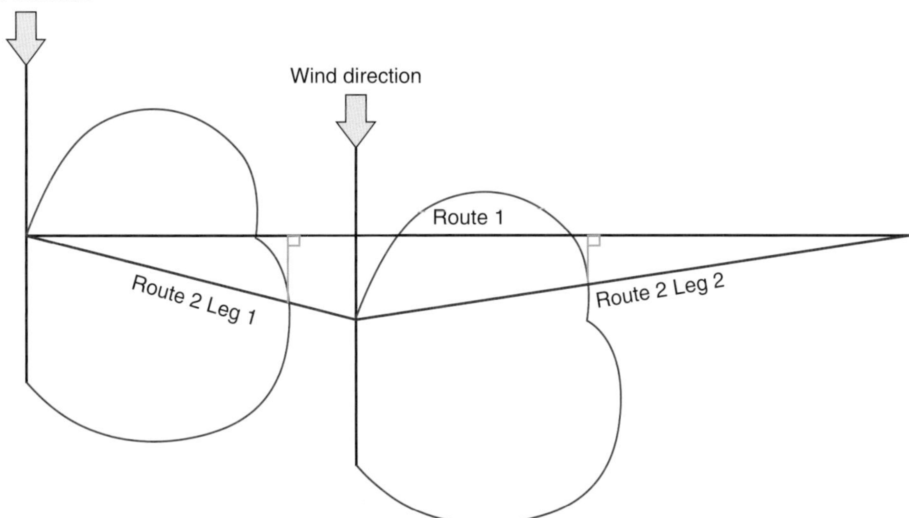

Route 1 – 14.4 nm at 4.1 knots
= 3 hours 30 minutes 44 seconds
Route 2 Leg 1 – 5.6 nm at 4.7 knots
Route 2 Leg 2 – 9 nm at 4.2 knots
= 3 hours 20 minutes 3 seconds

DIAGRAM 6.5 Using a Spinnaker for part of a Leg to Optimise VMC.

the spinnaker is up and areas where you are two-sail reaching. This acts in the same way as the concavity in the polar at upwind and downwind angles. By sailing as high and as fast as you can with the spinnaker, and then dropping it and going as fast as you can with two-sails up, you are effectively 'tacking' down the reach – optimising your VMG just as you would upwind or downwind. It is a technique well established in the dinghy classes when you cannot lay a gybe mark with the spinnaker up, so you sail high and fast until the angle is such that you can hold the spinnaker and then hoist. Once it is up, you can hold it round the mark and onto the next reach, following the same principle when dropping it to make the leeward mark.

Things get even more interesting when we apply the principles of VMC to the upwind and downwind legs. Imagine the boat sailing towards the windward mark. The wind is blowing directly from the mark and there is no current; we can then see that the optimum VMG course is the same as the optimum VMC course (Diagram 6.6a). However, if the wind shifts we must rotate the polar table round to line up with the new wind direction. The optimum VMC and VMG courses no longer match – we can get to the windward mark faster by sailing at a different angle to the optimum VMG. If we rotate the wind so that it has lifted us on this tack, then we see that we should sail lower and faster to optimise our VMC to the mark (Diagram 6.6b), and if it heads us then we should sail higher and slower (Diagram 6.6c). If you are going to use this technique you will need a very accurate polar table, right down to the tenths-of-a-knot that each couple of degrees of change in wind angle will give you – something that we have already written off as pretty much impossible.

That's not quite the case – and one place where the resources are available to make it possible, is the America's Cup. It was at the 1987 America's Cup in Fremantle, Perth, where this technique was developed by the Stars and Stripes team. Not only did they have the time and resources to refine their polar tables to the necessary extent, but they were also blessed with remarkably consistent conditions of wind gradient and sheer. The Fremantle Doctor provided the perfect opportunity for this to work, allowing them to sail to a different target speed, depending not only on the wind speed, but also on the wind direction. So concerned were they that this should not fall into the hands of the other syndicates that there was considerable resistance to the placement of TV cameras on board Stars and Stripes. After finally agreeing, they developed a code name for the information to hide its significance. Since the Fremantle Cup every challenger for the America's Cup has invested millions of dollars in on-the-water testing and polar development to back up the design theory.

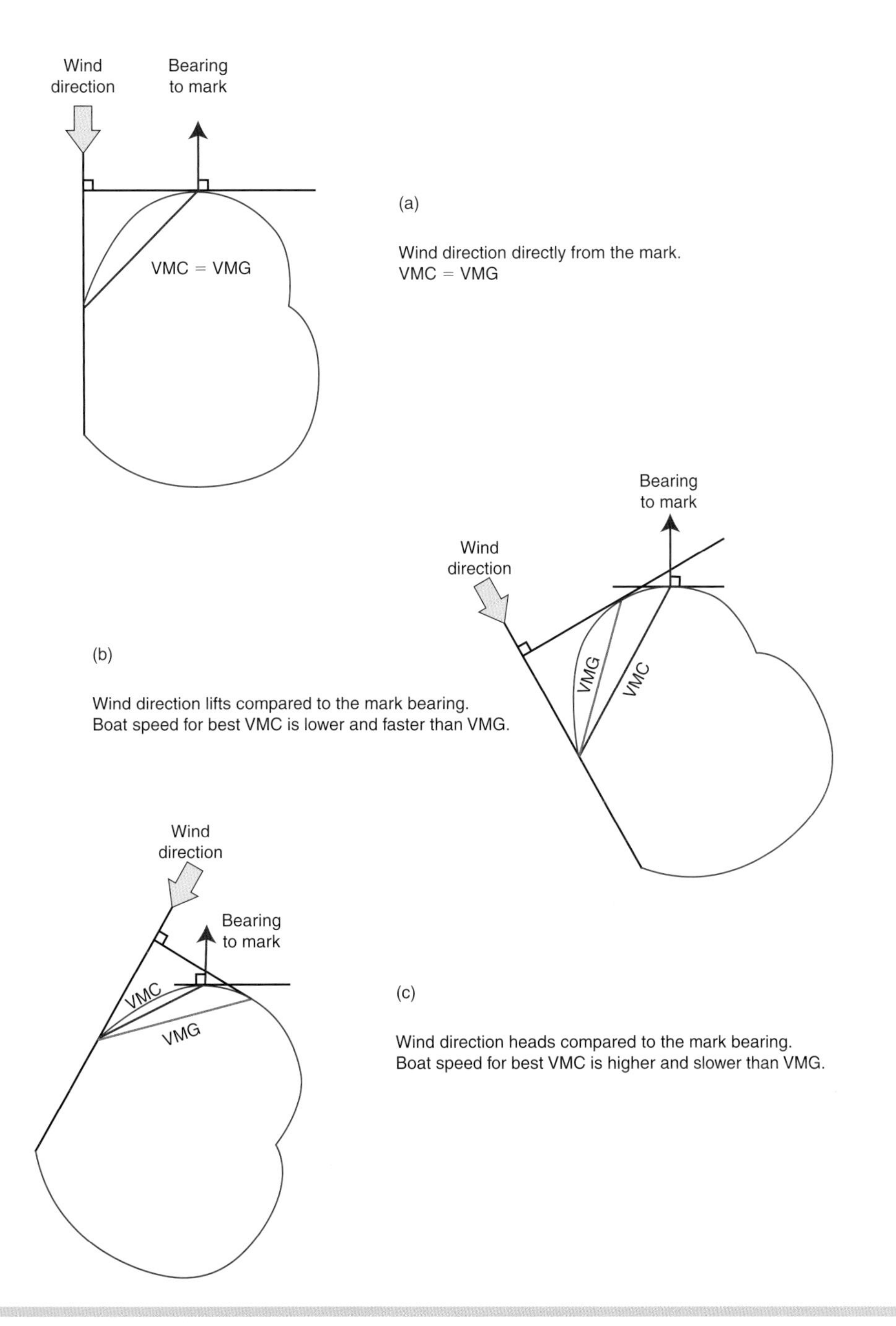

DIAGRAM 6.6 Sailing a VMC Course Upwind in Wind Shifts.

For the rest of us, without the resources of an America's Cup syndicate or such consistent wind conditions, developing the polar to that extent is, unfortunately, just about impossible. Nevertheless, the general rules can be applied; sail fast in the lifts and high in the headers. Not by much, perhaps just a couple of tenths on the boat speed dial. You will often find that this agrees with more general tactical rules. If you have got out to one side of the course and the fleet, when you get the header to tack back and consolidate you need to sail and cross, as fast as possible, as many boats as you can, so you sail fast on the lift that takes you across the fleet. Similarly, if you are trapped on a header by a boat to windward that will not tack you can minimise your losses by holding high and slow and letting him sail over you quickly so that you can tack.

WEATHER AND CURRENT ROUTING

Weather and current or tidal routing are the ultimate strategic tools, the one way that you can balance the most complex patterns of wind and water flow and come up with a definitive fastest route. There has to be a catch and there is – you need to know what the wind and current will do in advance. In the case of the tide this is not so bad, and tidal routing is really important in races where it is well understood – such as the English Channel. The wind is more difficult, whether you are tracking large pressure systems across the Atlantic, or trying to predict a wind bend up the next leg of an Olympic triangle, the weather can be equally uncooperative. Despite this, weather routing has scored some great successes.

So how does it work? If we look at Diagram 6.7 we know that the polar table describes the speed at which the yacht will sail at all wind angles, at any particular true wind speed. If we fix the time at one hour it will also show all the places that the yacht can reach in one hour. If, after the first hour the wind or the tide changes, then a second polar table rotated to account for the new conditions could be appended to the first at all points (or in practice every ten degrees of true wind angle). This would then show all the places the boat could reach in two hours. This technique can be repeated as often as you like, though doing it graphically would quickly become unmanageable.

This is where tactical software such as Deckman for Windows (or DfW) is so valuable. The software is designed to repeat the calculations until the expanding wave of 'places

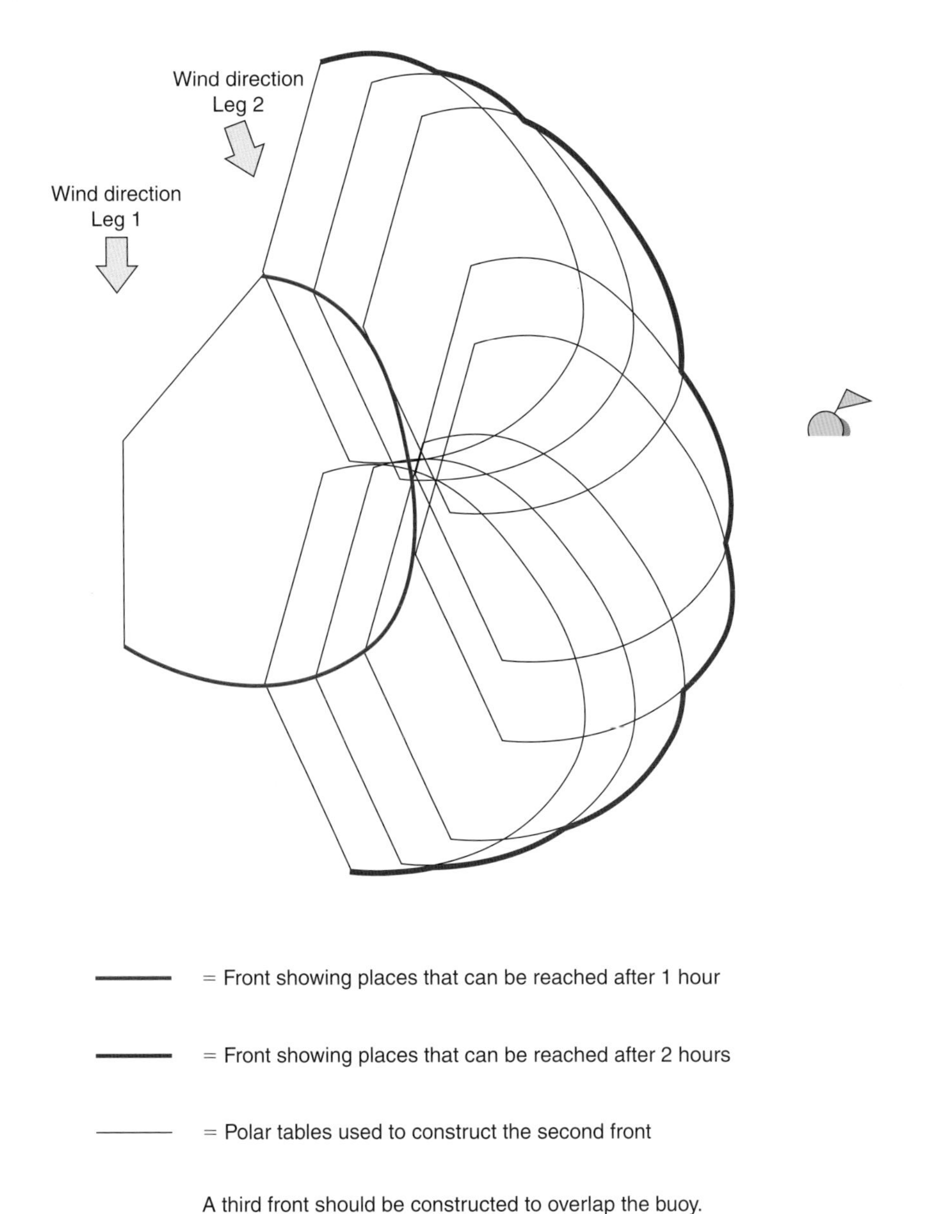

DIAGRAM 6.7 The use of the Polar Table in Weather Routing Conditions.

the yacht can reach in an equal period of time' overlaps the destination. Then, working back through the calculations, the optimum route to this point will be given by the angle sailed on each polar table 'leg'. If you follow the computer suggested route you will sail the fastest possible way from A to B – at least that is the theory.

We discussed the implication of polar table errors in the previous section, although in this instance it is a lesser problem than that of the wind and current forecasting. We have already mentioned the problems with the predictions of wind and tide so that the polar table can be set at the right angle for each leg. This is particularly acute for long legs such as those of the Volvo Ocean Race. The solution is only of any value if the front of places that you can reach in an equal period of time completely overlaps the destination. For an English Channel passage, or even a seven-day sprint across the Atlantic, forecasting to the finish is possible. But a thirty-day Volvo leg sees your weather predictions sliding from the reasonably accurate to the unreasonably hopeful. For the system to work well the operator must have a clear idea of these problems and how they may implicate his decision. The routing becomes a sophisticated 'what if' calculator, allowing you to set up several scenarios and see where they take you. Choosing your course may be a question of balancing the various possibilities and coming up with the lowest risk strategy.

In many ways, weather and tidal routing systems epitomise the way the job specification for a navigator has changed in the last few years. It is not necessarily any harder, but it is certainly different. The skills that were formerly involved in keeping the constant plot running on the chart (without a GPS) are just as difficult to master as the complexities of the modern instrument and computer system. The difference is that the navigator can now contribute a great deal more to the sailing of the yacht – both its performance analysis and the tactics. All the techniques we have looked at can be learned by anyone due to the emergence of commercially available systems with all the functionality of the best Grand Prix equipment. The calibration of the instruments, generation of the polar tables, and then the application of all this to navigational and tactical problems, defines one of the most complex roles on a modern race boat. I hope that this book has provided some insight to it.

A Quick Guide to Calibration

NOTE: The text below refers in places to B&G products and their features. Much of the information and techniques are applicable to other manufacturers' instruments and displays. Refer to specific product user guides for further information.

Depth

For obvious safety reasons depth should be calibrated before any other function. A DATUM (offset value) is set such that the depth display refers either to the waterline or bottom of the keel. Enter a positive offset for waterline or a negative offset for keel.

Heading

In order to eliminate compass deviation errors caused by magnetic fields in the yacht, heading calibration is carried out using B&G's AutoSwing. An offset is then entered to account for the alignment of the compass as installed on the yacht.

Prior to commencing the swing, ensure that no magnetic or large metallic objects are placed near to the compass. Choose a calm open stretch of water and then begin the calibration process as outlined in the User Manual. Ensure that you steer a steady circle not exceeding 2–3 degrees per second with a boat speed of less than 5 knots.

Having completed the swing, compare the compass heading with the bearing of a known transit and correct it by entering an offset value.

Boat Speed

AutoCal (Distance Reference)

All the work in the world on target speeds and angles or on polar tables goes to waste if boat speed is inaccurate, even by a few tenths of a knot.

True wind calculations and performance functions depend on this input being accurate. If boat speed is wrong, the whole system's data quality is compromised.

- It is important to keep an eye on boat speed accuracy during the course of a season as paddlewheels do suffer from friction and hence changes in physical characteristics can occur over time.
- Regular cleaning of the paddlewheel is essential for accurate and repeatable speed readings.

The B&G three run method eliminates any changes in current during the calibration procedure since the runs are carried out in alternate directions. When carrying out each run maintain a constant compass course to ensure that the distance covered is the same as that measured on the chart. If possible choose a measured distance with clearly visible transits marking each end.

It is prudent to manually calibrate boat speed using the trip log function on another GFD/FFD whilst running the AUTO CAL feature. By recording the log at the beginning of each run, and recording the log distance run for each leg, we can deduce the error in current speed calibration. This can be useful to complete a calibration procedure if any of the runs have to be aborted due to a necessary change of course.

Example:

- Current calibration value in Hz per Kt = 3.42
- Measured run distance = 0.75 nm
- Run 1 = 0.76 nm, Run 2 = 0.72 nm, Run 3 = 0.77 nm

It appears that the current is changing due to the difference in logged distance for the 1st and 3rd runs. Correct this by averaging run 1 and 3, and then combine with run 2 to get an overall average log distance:

$$(\text{Run1} + \text{Run3}) \div 2 = 0.765$$

$$(\text{Run2} + 0.765) \div 2 = 0.7425$$

Comparing the log distance to the actual distance reveals that the log is under-reading by 1%

$$(0.75 - 0.7425) = 1\%$$

Increasing SINGLE HZ calibration value in manual calibration reduces boat speed, and vice versa. Therefore by reducing the SINGLE HZ value for the above calibration by 1% will address the under reading of the log.

$$3.42 - 1\% = 3.39$$

If you are using the latest B&G Hercules system with a heel sensor you can also adjust the speed from tack to tack and apply more correction at higher speeds in the case of the paddlewheel. See the Owner's Manual for more information on this.

Once boat speed has been verified as accurate, it should be monitored against other functions, and only adjusted if the wrong speed cannot be attributed to different sail trim, helmsman, rig settings, wind shear, crew weight and hull fouling.

AutoCal (SOG Reference)

The H3000 provides an Auto Cal facility that uses SOG from your GPS and compares the average of this against the average boat speed from the speed sensor for the duration of the calibration run. Note: This will only work accurately in non-tidal conditions.

Apparent Wind Angle

Installation offsets and other asymmetric factors can affect the symmetry of measured wind angles. An offset is all that is required to correct for tack-to-tack errors in APP W/A.

The perfect scenario for calibrating APP W/A is in non-tidal flat water, with stable conditions (no building sea breeze) and with a correctly calibrated boat speed.

Using identical trim and crew weight distribution (not necessarily a full racing crew) complete half a dozen tacks with the helm concentrating purely on boat speed. A consistent speed (not necessarily from the targets table) on each tack should be achieved without changing trim, and without using any reference to APP W/A. The navigator or tactician should monitor the APP W/A from tack to tack, and average each tack. Any difference between Port and Starboard tack can be corrected in the MHU offset.

If Starboard APP W/A is greater than Port then subtract half the difference from the MHU offset, and vice versa.

Once a good APP W/A is achieved, do not change it unless necessary (e.g. rig has been removed from boat, new wind sensor put in place). Any differences seen from now on may be attributable to effects such as wind shear and current.

Apparent Wind Speed

Do not alter the APP W/S calibration values. This is a wind tunnel based factory setting.

True Wind Angle Calibration

It is usual to start the TWA calibration process by setting the boat up to do a number of tacks upwind in as steady conditions as possible.

A TRUE W/A calibration matrix is utilised for corrections upwind, reaching, and downwind across the TRUE W/S ranges. Two methods of calibrating TRUE W/A are available, either monitoring TRUE DIR from tack to tack or gybe to gybe, or using the compass to verify the angles the yacht is tacking or gybing through.

If an error is seen in TRUE DIR, then the following rule applies:

- If TRUE DIR is lifting you tack to tack then TRUE W/A is reading too wide, half the error must be subtracted from the correction table.
- If TRUE DIR is heading you tack to tack then TRUE W/A is reading too narrow, add half the error to the TRUE W/A table.

If according to the compass you are tacking through an angle different than the sum of the TRUE W/A's on each tack (Port TRUE W/A + Starboard TRUE W/A) then the following rule applies

- If the tack angle < the sum of the TRUE W/A's, the TRUE W/A is reading too wide, half the error must be subtracted from the correction table.
- If the tack angle > the sum of the TRUE W/A's, the TRUE W/A is reading too narrow, add half the error to the TRUE W/A table.

A Worked Example:

Your TRUE DIR is telling you that you are being lifted by 10 degrees from tack to tack (or the sum of your TRUE W/A is 10 degrees more than the angle you are tacking through), and you know this is wrong. This is in 5 kts of TWS. When you go into the TRUE W/A correction table for 5 Kt upwind, you see that a value of −1.5 degrees is already entered.

To correct the error you are seeing, a further 5 degrees needs to be subtracted so a total correction of −6.5 degrees is applied.

Calibrating and refining TRUE W/A is an ongoing process. Differences in TRUE W/A can be seen from morning to afternoon races as sea breezes develop and gradient conditions change. TRUE W/A should be checked at the beginning of every race, any necessary changes made, and the conditions logged for future reference.

True Wind Speed Calibration

TRUE W/S errors are seen from sailing upwind to downwind. This is most noticeable on Mast Head boats, however all yachts are affected to some degree. This is due to the acceleration of the airflow over the top of the mast and around the sails when sailing downwind. The introduction of the Vertical Mast Head unit has gone some way to solving this, however calibration may still be necessary.

As a rule of thumb, it is safe to put a correction of −10% into the table before first sailing and this is the value entered by B&G as a factory default. Monitoring the change in TRUE W/S from close hauled to flat running will enable further refinement of this calibration value. On a B&G Hercules system it is also possible to define the TRUE W/A at which this correction is applied – typically your downwind target angle.

Example

	True Wind Speed (kts)					
	5	10	15	20	25	30
Upwind TWA Correction	−7.0	−3.0	2.0	5.0	6.5	8.0
Reaching TWA Correction	−6.0	−2.0	0.0	1.0	1.0	1.5
Downwind TWA Correction	4.0	3.0	1.0	−1.0	−1.0	−2.0
Downwind TWS Correction	−1.5	−1.8	−2.5	−3.0	−3.5	−4.0

General Maintenance

Through-Hull Housings

Keep the screw threads of through-hull housings well greased with silicone or water pump grease. Ensure that the outer surfaces of the housing are properly coated with anti-fouling paint.

Boat Speed Sensor (paddlewheel type)

Use a stiff brush to remove marine growth that may cause the paddlewheel to freeze, and then clean the surfaces with a very weak solution of household detergent. If fouling is very severe, push the paddlewheel axle out by using a small drift, and then very gently, wet sand the surface with a fine grade wet/dry paper.

Inspect the o-rings on both the sensor and the blanking plug and replace if necessary, and then lubricate with silicone lubricant or petroleum jelly (Vaseline®).

Boat Speed Sensor (sonic type)

Aquatic growth can accumulate rapidly on the transducer surface reducing performance. Clean the surface with a soft cloth and a very weak solution of household detergent.

If fouling is severe, use a stiff brush or a putty knife. Take care not to cause scratches on the transducer face. Wet sanding using fine grade wet/dry paper is permissible to remove stubborn deposits.

Surfaces exposed to salt water must be coated with antifouling paint. Use only water-based antifouling paint. Solvent-based paints must not be used. Solvent based paints contain 'ketones' which may attack the plastic surfaces and damage the sensor. Re-apply the antifouling paint every six months or at the start of each boating season.

Desiccators

Should any display window show signs of moisture having penetrated the seals e.g. misting of the glass or condensation, the instrument should be removed and returned to your national distributor for drying.

Masthead Unit

Storage of the masthead unit when the yacht is laid up afloat will increase the life of the transmitters. It should always be removed from the masthead before the mast is unstepped. It should be stored in its packing box with the vane and cups removed. The exposed socket and connector threads at the top of the mast should be smeared with silicone grease such as MS4 (Midland Silicones Ltd), and then protected with the plastic cap supplied with it.

The contacts in the masthead unit connector should be inspected for cleanliness and sprayed with a water inhibitor such as WD40. The outer casing of the connector should also be smeared with silicone grease.

The masthead unit must never be oiled. The bearings are of the sealed pre-lubricated type and any additional oil may cause chemical breakdown of the existing lubricant. Any scratch marks or corrosion on masthead unit spar should be rubbed clean with a soft cloth and lightly smeared with silicone grease. This should not be necessary if care is taken when hoisting or lowering the masthead unit, to protect it from collision against the rigging.

If the mast is un-stepped, care must be taken to ensure that the cable is not cut through, but disconnected at the junction box below decks. The bare ends of the cable should be smeared with silicone grease.

Calibration Records

NOTE: Some of the tables below refer to B&G products and their features. Many aspects of these tables are applicable to other manufacturers' instruments and displays. Refer to specific product user guides for further information.

H3000 SYSTEM CALIBRATION RECORD

System Configuration Record

Function	Default Setting	User Setting
Heading Node	16 (Halycon 2000)	
Halycon Node	0	
Linear 1	4 (Heel)	
Linear 2	5 (Trim)	
Linear 3	6 (Barometer)	
Linear 4	1 (0–1000 Type)	

(*Continued*)

(Continued)

Function	Default Setting	User Setting
NMEA Channel	0	
Baud Rate	6	
Sea Temp Type	1	

Basic Calibration Record

Function	Calibration	Value
Meas W/A	Offset	
Meas W/S	Hz/Kt	1.04 (default)
	Offset	1.04 (default)
App W/A	Offset	
	Heel Correction	
App W/S	Hz/Kt	1.04 (default)
	Offset	1.04 (default)
Boat Speed	Single Hz/Kt	
	Stbd Hz/Kt	
	Port Hz/Kt	
Heading	Offset	
Depth	Datum	
Heel	Offset	
Trim	Offset	
Leeway	Coefficient	
Mast Angle	Offset	
Mast Height	Mast Height	15.0 Metres (default)
Rudder	Offset	

True Wind Speed Correction Table

Function	True Wind Speed (kt)					
	5	10	15	20	25	30
Correction °						
Correction Angle						

True Wind Angle Correction Table

Wind Speed	True Wind Speed (kt)					
	5	10	15	20	25	30
Upwind						
Reaching						
Downwind						

Boat Speed Correction Table

Heel Angle	Boat Speed (kt)					
	5	10	15	20	25	30
0°						
10°						
20°						

Damping Record

Function	Calibration	Value
App W/A		
App W/S		
Heading		
Boat Speed		
Heel		N/A
Trim		N/A
Leeway		N/A
Mast Angle		N/A
Rudder		N/A
True W/A		
True W/S		
True Dir		N/A
Tide		N/A